CW00558586

tudent Support
Materials for

AQA

.-level Year 2

Biology

**opics 7 and 8: Genetics, populations, evolution
nd ecosystems, The control of gene expression**

uthor: Mike Boyle

William Collins' dream of knowledge for all began with the publication of his first book in 1819.

A self-educated mill worker, he not only enriched millions of lives, but also founded a flourishing publishing house. Today, staying true to this spirit, Collins books are packed with inspiration, innovation and practical expertise. They place you at the centre of a world of possibility and give you exactly what you need to explore it.

Collins. Freedom to teach

HarperCollins Publishers
The News Building
1 London Bridge Street
London SE1 9GF

> **Browse the complete Collins catalogue at**
> **www.collins.co.uk**

10 9 8 7 6 5 4 3 2 1

© HarperCollins*Publishers* 2016

ISBN 978-0-00-818948-8

Collins® is a registered trademark of HarperCollins*Publishers* Limited

www.collins.co.uk

A catalogue record for this book is available from the British Library

Commissioned by Gillian Lindsey
Edited by Alexander Rutherford
Project managed by Maheswari PonSaravanan at Jouve
Development by Kate Redmond and Gillian Lindsey
Copyedited by Rebecca Ramsden
Proof read by Janette Schubert
Original design by Newgen Imaging
Typeset by Jouve India Private Limited
Cover design by Angela English
Printed by CPI Group (UK) Ltd, Croydon, CR0 4YY
Cover image © iStock/royaltystockphoto

All rights reserved. No part of this book may be reproduced, stored in a retrieval system, or transmitted in any form or by any means, electronic, mechanical, photocopying, recording or otherwise, without the prior permission in writing of the Publisher. This book is sold subject to the conditions that it shall not, by way of trade or otherwise, be lent, re-sold, hired out or otherwise circulated without the Publisher's prior consent in any form of binding or cover other than that in which it is published and without a similar condition including this condition being imposed on the subsequent purchaser.

HarperCollins does not warrant that www.collins.co.uk or any other website mentioned in this title will be provided uninterrupted, that any website will be error free, that defects will be corrected, or that the website or the server that makes it available are free of viruses or bugs. For full terms and conditions please refer to the site terms provided on the website.

Contents

3.7 Genetics, populations, evolution and ecosystems

3.7.1 Inheritance

There's an important difference between having genes and **expressing** them. Humans and chimps may possess very similar genes but some remain unused while others are expressed. A gene can only have an impact on an organism if it is switched on and used to make a particular protein.

Why are we all different? Let's start with two important definitions. The **genotype** of an organism is the genetic makeup, the collection of **alleles** that the organism has inherited from its parents. The **phenotype** of an organism is the collection of observable features that it has.

Why do we turn out the way that we do? There are two possible causes of variation, sometimes referred to as 'nature v nurture'. Is it the genes we have inherited (nature), or are environmental factors important (nurture)? The answer is usually that it's a complex interaction of both.

Essential Notes

Put simply: genotype + environment = phenotype

Definitions

Genetics: some key definitions

Genotype – the genetic constitution of an organism. When we study the inheritance of one gene we use the term genotype to describe the combination of **alleles** the individual possesses. This is written as, for example, BB or Bb.

Phenotype – the observable features of an organism. The phenotype results from a combination of the genes the organism inherits and the environmental factors. For example, we are all born with a certain combination of tallness genes so we have the potential to grow to a particular height. However, we will not reach this height without an adequate diet.

Gene – a length of DNA that codes for the manufacture of a particular protein or polypeptide. In this section, think of it as a piece of DNA that codes for a certain feature. Humans, like all other animals, are **diploid** organisms – we have two copies of each gene.

Chromosome – one long DNA molecule with genes dotted along its length (Fig 1).

Locus – the position of a gene on a chromosome.

Allele – an alternative form of a gene. Genes often have two or more alleles; for example, a flower colour gene could have two alleles, such as one for red flowers and one for white flowers.

> **Dominant** – the allele that, if present, is shown in the phenotype; for example, the allele B could cause black fur in mice.
>
> **Recessive** – the allele that is only expressed in the absence of the dominant version; for example, the allele b could code for brown fur in mice.
>
> **Homozygous** – when both alleles are the same: BB or bb.
>
> **Heterozygous** – when a pair of alleles is different: Bb.

When an egg is fertilised a single set of **chromosomes** (Fig 1) from the male matches with a single set from the female, so there are two sets of chromosomes. In these **diploid** cells there are two copies of each gene – one allele from the mother and one from the father.

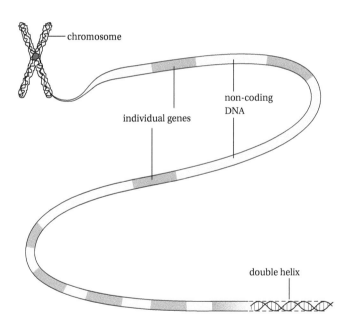

Fig 1
Chromosomes contain many genes that are lengths of DNA that code for a particular protein.

Mendel's genetics – monohybrid and dihybrid inheritance

The basic rules of inheritance were worked out in the 1860s by Gregor Mendel. However, his work remained largely unappreciated until around 1900, when the process of **meiosis** was first observed and understood. Meiosis (see *Collins Student Support Materials AS/A-Level year 1 - Topics 3 and 4*) went a long way to explaining Mendel's observations.

Table 1
Possible genotypes for mice with coat colour B and b

Monohybrid inheritance

This refers to the inheritance of a single gene with two alleles. For example, coat colour in a particular strain of mice is determined by one coat colour gene with two alleles, B (for black fur) and b (for brown fur).

There are three possible genotypes for these mice, but only two phenotypes, as shown in Table 1.

Genotype	Phenotype
BB (homozygous dominant)	Black
Bb (heterozygous)	Black
bb (homozygous recessive)	Brown

Notes

Every year candidates lose marks by writing ratios in the wrong way. A ratio of 1:3 is not the same as 1 in 3; the ratio of 1:3 is the same as 1 in 4, and can also be expressed as 0.25 or 25%.

Fig 2
Monohybrid inheritance in mice. Pure breeding black and brown mice are homozygous (BB and bb). When bred, the mice born in the first litter are all black and are heterozygous (Bb); if these mice are then bred together, the litters have black and brown mice in a ratio of 3:1.

Notes

In questions about genetic crosses, there are several different ways of expressing the outcome.

For example 25%, 0.25, 1 in 4 and a 3:1 phenotypic ratio all refer to the same thing. By convention, probabilities should be expressed in decimals, so if a probability is asked for, 0.25 is the best answer.

Note that black is **dominant** because if the allele B is present, it is shown in the phenotype. The b allele is **recessive**, and is only expressed in the absence of the other allele. Fig 2 shows the basics of a **monohybrid cross**. If a Bb mouse mates with a Bb mouse, there is a three-to-one ratio in the phenotypes of the offspring (a **phenotypic ratio**).

In genetics we deal with large numbers of **gametes**, so the ratios are averages. It's not like the draw for the FA Cup, where you know what's been drawn and therefore what's left in the bag. If the chances of a child being born with genotype bb is 1 in 4, and the couple have three children of genotype BB, Bb and Bb, then the chance that the fourth will be bb is still 1 in 4.

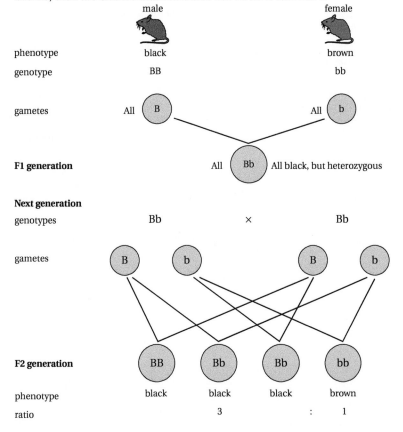

Codominance

There are cases where alleles are **codominant**, so if they are present they are both expressed in the phenotype. A classic example is the human ABO blood group system (which is also an example of a **multiple allele**) where there are more than two alleles of the same gene. Some genes have many different alleles.

The ABO system is controlled by one gene, I, with three alleles, I^A, I^B and I^O. I^A and I^B are codominant over I^O.

- Allele I^A codes for A proteins on the red cells.
- Allele I^B codes for B proteins on the red cells.
- Allele I^O codes for no relevant proteins on the red cells.

This gives us the table of genotypes and phenotypes shown in Table 2.

Genotype	Phenotype (i.e. blood group) and explanation
I^O, I^O	O – I^O is recessive
$I^A I^O$ or $I^A I^A$	A – I^A is dominant over I^O
$I^B I^O$ or $I^B I^B$	B – I^B is dominant over I^O
$I^A I^B$	AB – I^A and I^B are codominant

Table 2
The genotypes of blood groups

Dihybrid crosses

A **dihybrid cross** involves two separate genes, each with two alleles, at the same time. It is assumed that the genes are located on different chromosomes so that they can be separated by meiosis.

The vital steps in working out the products of a dihybrid cross (Fig 3) are:

1 Parents' genotypes.

2 Gamete formation – this is where you can apply your knowledge of meiosis. One allele from each pair can go into the gamete, but by independent segregation it can pass into the gamete with any allele from the other pair. Thus a genotype of AaBb can give gametes of AB, Ab, aB and ab.

3 By random fertilisation any male gamete can combine with any female gamete.

4 With dihybrid inheritance there can be up to 16 different genotypes, so it is important to organise the results using a **Punnett square**.

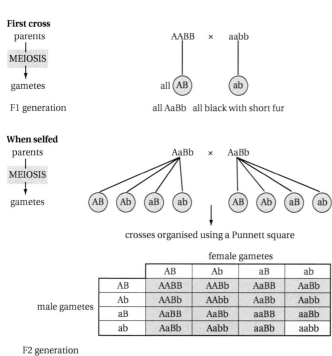

Fig 3
An example of dihybrid inheritance – a cross involving two genes. We have already looked at the mouse coat colour gene, where the allele B for black hair is dominant to the allele b for brown hair. A second gene controls hair length. Allele A for short hair is dominant to allele a for long hair. A cross between two pure-breeding mice, one black with short hair and the other brown with long hair, will produce a litter of all-black mice. However, these mice are all heterozygous for both genes. If these mice are bred together to produce a second generation (F2), all possible combinations are produced in the classic ratio of 9:3:3:1.

First cross

parents   AABB   ×   aabb

MEIOSIS

gametes   all (AB)   (ab)

F1 generation   all AaBb   all black with short fur

When selfed

parents   AaBb   ×   AaBb

MEIOSIS

gametes   (AB) (Ab) (aB) (ab)   (AB) (Ab) (aB) (ab)

crosses organised using a Punnett square

female gametes

male gametes		AB	Ab	aB	ab
	AB	AABB	AABb	AaBB	AaBb
	Ab	AABb	AAbb	AaBb	Aabb
	aB	AaBB	AaBb	aaBB	aaBb
	ab	AaBb	Aabb	aaBb	aabb

F2 generation

Black, short fur	Black, long fur	brown, short fur	Brown, long fur
9	3	3	1

Essential Notes

A key point here is that, with sex-linked inheritance, males can't be carriers.

Sex linkage

Sex-linked inheritance concerns genes found on the sex chromosomes, in humans these are the X and Y chromosomes. In humans, the X chromosome is larger and contains more genes, so most examples of sex linkage in humans concern X-linked genes. Y-linked traits are known, but are rare.

With sex-linked inheritance, the normal pattern is affected by the key fact: *males only have one X chromosome; so have only one copy of a sex-linked allele.*

Haemophilia is an example of a sex-linked disease. This disease is caused by a faulty allele that fails to make the correct blood clotting protein, **factor VIII**. There are two alleles: the normal allele, H and the faulty allele, h. In sex linkage we show the chromosomes as well as the alleles.

There are three possible genotypes for females:

- $X^H X^H$ = a healthy female; normal alleles on both X chromosomes
- $X^H X^h$ = a healthy female, but a carrier of the haemophilia allele
- $X^h X^h$ = a haemophiliac (very rare in females).

Males have only two possible genotypes:

- $X^H Y^-$ = a healthy male; there is no allele on the Y chromosome
- $X^h Y^-$ = a haemophiliac; he has no healthy gene on the Y chromosome to mask the effects of the faulty gene he has inherited

Autosomal linkage

When two genes are found together on the same **autosome**, they are said to be linked. The vital point here is that linked alleles cannot be separated by independent segregation, so will always be inherited together unless separated during **crossover**.

From this definition it would seem that linkage is a black and white issue; if they are on the same chromosome, they are linked. However, in practice, there are degrees of linkage. Genes whose loci are close together on the chromosome will almost always be inherited together because the chances are slim that **chiasmata** will form between them, allowing crossover. Genes whose loci are on opposite ends of the chromosome will almost always be separated by crossing over. You can tell a lot about the position of different loci by how often they are separated by crossing over.

Example of autosomal linkage

In the sweet pea plant, the allele for purple flowers is dominant to the allele for red flowers, and the allele for elongated pollen is dominant to that for round pollen.

If pure-breeding (in other words, homozygous) plants with purple flowers and elongated pollen grains are crossed with plants with red flowers and round pollen grains, the first (F1) generation are all purple with elongated pollen grains. When the F1 offspring are selfed (bred together), you would expect a 9 : 3 : 3 : 1 phenotypic ratio in the next generation. This does not happen, because the genes for flower colour and pollen grains are found on the same chromosome. There are more plants like the 'parents' – purple/elongated and red/round – and fewer recombinants (offspring that have different allele combinations to their parents). The actual results are shown in Table 3.

Phenotype	Approximate numbers expected if no linkage (9 : 3 : 3 : 1 ratio)	Observed numbers
Purple, elongated	405	336
Purple, rounded	135	31
Red, elongated	135	28
Red, rounded	45	325
Total	720	720

Table 3
F2 phenotypes for a dihybrid cross

Explanation

The purple/rounded and red/elongated plants are recombinants, and must be the result of crossover. Of the 720 offspring, 59 (31 + 29) are recombinants. This is 8% and so we can say that the loci for flower colour and pollen shape are eight genetic units apart. If they were further apart, there would be more opportunities for chiasmata to form between them, and there would be more recombinants.

Epistasis

Epistasis is a situation where the expression of one allele depends on the presence of a particular allele of another gene.

For example, the natural colouration of wild mice is called agouti, and is caused by 'banded' hairs which have a dark base and a lighter top.

There are two genes involved, one for the pigment and one for the bands.

- Allele A codes for the ability to make pigment. This is dominant.
- Allele a codes for the inability to make pigment.
- Allele B codes for the ability to make banded hairs. This is dominant.
- Allele b codes for no bands.

From this it can be see that, in order to produce the agouti colouration, you need the ability to make bands *and* the ability to make pigment, so all genotypes that include alleles A and B will be banded (see Table 4). This is written as A-B-.

In the absence of allele A, allele B cannot be expressed, so all genotypes of aa will be albino. All genotypes of A-bb will be black because the individual can make pigment but not bands.

Genotype	Diagram	Phenotype	Explanation
AABB AABb AaBb AaBB		Agouti; mice with banded hairs	These mice possess both alleles A and B so can produce pigment and bands
AAbb Aabb		Black mice	Can make pigment but not bands
aaBB aaBb aabb		Albino mice	Do not possess allele A so can't make pigment. B/b becomes irrelevant.

Table 4

The chi-squared test in genetics

The **chi-squared (χ^2) test** is a statistical test that can be used to see whether there is a significant difference between the actual results obtained (the 'observed') and those you would expect given your understanding of the particular genetic cross.

When a particular cross is made and the phenotypic ratios of offspring are not as expected, the chi-squared test will give us a probability that the results are simply due to chance or if there some other interaction at work, such as epistasis or linkage.

Example – chi-squared test in genetics

A yellow-flowered plant was crossed with a red-flowered plant. The resulting offspring (F1 generation) were all orange. This would suggest a straightforward hypothesis: there is one gene with two alleles. Allele R codes for red and r codes for yellow flowers. It seems that these alleles are codominant, and a genotype of Rr results in orange flowers. If this hypothesis is true, you would expect, from probability, a ratio of 1 red : 2 orange : 1 yellow in the F2 generation. Table 5 shows the results for a self-breeding cross of the F1 generation.

Table 5
F2 phenotypes for a monohybrid cross

Phenotype	Observed numbers	Expected numbers
Red	77	80
Orange	172	160
Yellow	71	80
Total	320	320

There were 320 individual plants in this F2 generation, which is more than enough to minimise sampling error, so you would expect something close to 80 : 160 : 80.

So how well is the hypothesis supported by the observed results? Generally, statistical tests can only be done on a **null hypothesis**: so we need a negative version of our hypothesis such as 'the results are not due to simple codominance, some other mechanism is operating'.

The chi-squared test can be used to find out whether or not we can reject our null hypothesis. It's an 'innocent until proven guilty' approach. Chi-squared, χ^2, is always calculated from actual numbers and not percentages, proportions or fractions. A good sample size is therefore very important.

The formula for chi-squared, χ^2, is:

$$\chi^2 = \sum \frac{(O - E)^2}{E}$$

where O represents the observed results and E represents the results that we should get using the expected ratio from our hypothesis.

Table 6 shows the results of the chi-squared test.

Phenotype	Observed (O)	Expected (E)	(O–E)^2	(O–E)^2/E
Orange	172	160	144	0.9
Yellow	71	80	81	1.01
Red	77	80	9	0.1
				χ^2 = 2.01

Table 6
Working out chi-squared

We now need to look up what probability this value for χ^2 relates to (Table 7). We really want to know whether our χ^2 value is above or below 0.05. A probability of less than 0.05 means a significant difference between observed and expected results.

We must first work out the number of **degrees of freedom**. The number of degrees of freedom is one less than the number of categories (the number of possible phenotypes). This investigation has three categories so the number of degrees of freedom is two.

Degrees of freedom	Probability (p)							
	0.975	0.9	0.50	0.25	0.10	0.05	0.02	0.01
1	0.000	0.016	0.45	1.32	2.71	3.84	5.41	6.64
2	0.051	0.211	1.39	2.77	4.61	5.99	7.82	9.21
3	0.216	0.584	2.37	4.11	6.25	7.82	9.84	11.34
4	0.484	1.064	3.36	5.39	7.78	9.49	11.67	13.28

Table 7
Table of chi-squared values

Essential Notes

The outcome of many statistical tests is a value of p, probability. The value of p will be somewhere between 0 and 1. For example, p = 0.1 means that the probability that the results are due to chance is 0.1, or 10%, which means that you would expect to get results such as this by chance one in every ten times.

The usual significance threshold for p is 0.05, usually written as p ≥0.05, meaning that the probability that the results are due to chance is equal to or less than 0.05, or 5%. We can therefore be more than 95% certain that the results are significant.

For χ^2 = 2.01 and 2 degrees of freedom, the probability for our hypothesis is between 0.5 (50%) and 0.25 (25%) – highlighted in the table. This means that you would expect results to differ by as much as this between 50% and 25% of the time. In plain English, it is highly likely that these results fit in with our expected pattern – we don't need to look for another explanation. We can reject the null hypothesis, and accept the experimental hypothesis that it's a simple codominant mechanism operating.

3.7.2 Populations

The Hardy-Weinberg principle

In this section we look at populations rather than individuals.

A **Mendelian population** or **deme** is a population of organisms of one species that can actively interbreed with one another and share a distinct **gene pool**.

Imagine a gene for coat colour in a particular population of rats. Allele A is the dominant allele, producing brown fur. Recessive allele a produces albino rats. Rats of genotype AA and Aa are brown, while aa rats are albinos (they have white fur due to an absence of pigment).

It's easy to find out how many rats are genotype aa because we can see them. But what if we want to know how many are Aa? They look just the same as AA rats. We can use the **Hardy-Weinberg** equation to work out how many rats are carrying the recessive allele.

- If we use the letter p to denote the frequency of the A allele.
- And q for the frequency of the a allele.
- We know that $p + q = 1$ (1 is the same as 100%. There are only two possible alleles so when added up they must equal 100%.)

Each individual has two alleles.

- The frequency of AA rats is p^2.
- The frequency of Aa rats is $2pq$ (because the rats can be aA or Aa).
- The frequency of aa rats is q^2.

And as these are the only three possibilities:

$$p^2 + 2pq + q^2 = 1$$

In words, this formula says that the frequency of AA rats, when added to those that are Aa, aA, or aa, must equal 100%.

Worked Example

If 4% of the rats are albino:

(a) What is the frequency of the a allele?

(b) What proportion of the rat population is heterozygous, i.e. has the genotype Aa and so carries the a allele?

Answers:

(a) If 4% of the rats are albino, and therefore aa, we know that $q^2 = 0.04$.

Then the frequency of the a allele is $\sqrt{0.04}$, which is 0.2 or 20%.

We know that $p + q = 1$, so the frequency of the A allele is 0.8, or 80%.

(b) The question is asking 'how many rats are Aa?', which is $2pq$.
$2pq = 2 \times 0.8 \times 0.2 = 0.32$ or 32%.

Overall, from the observation that 4% of the rats are albino the Hardy-Weinberg equation tells us that 32% of the rats are Aa and 64% are AA.

In population genetics the Hardy-Weinberg principle states that the genotype frequencies in a population *remain constant* or are in equilibrium from generation to generation unless *specific disturbing influences* are introduced. These include the following:

- **Non-random mating** – for example, if brown rats tended to choose other brown rats as mates, or albinos only mated with other albinos.

- **Mutations** – so that new alleles were created, more than just A and a.

- **Selection** – where one coat colour gives the individual an advantage over other individuals with the other coat colour.

- **Limited population size** – when numbers are small, chance becomes significant.

- **Gene flow** – immigration or emigration from the population.

3.7.3 Evolution may lead to speciation

Natural selection

This idea is probably the most important in biology. Darwin's idea can be summed up as follows:

- There is always variation within a population.

- There will be competition for resources.

- Those individuals with the most favourable alleles, or combinations of alleles, will pass more of these alleles on to the next generation.

- The frequency of the favourable alleles will increase from generation to generation.

The result of natural selection is that individuals are better adapted to their environment. These adaptations can be:

- **anatomical** – such as hairs on leaves to reduce transpiration

- **physiological** – such as the possession of an enzyme that allows the individual to break down a particular pollutant or antibiotic

- **behavioural** – such as the instinct to migrate to avoid harsh conditions, or huddling to keep warm.

Natural selection is to do with **reproductive success**. Organisms that are best adapted to their environment will have a higher probability of passing on more of their genes to the next generation than will less well adapted organisms.

Definition

Natural selection is a process in which those organisms whose alleles or allele combinations give them a selective advantage are more likely to survive, reproduce and pass on their alleles to the next generation.

Natural selection is often thought of as just a mechanism for change, but it can also be a force that keeps things stable. There are three types of selection. (These are shown in Fig 4 and described below.)

1 **Stabilising selection** – this can lead to a standardising of organisms by selecting against extremes, especially in a stable environment. Birth weight of mammals is a good example. An abnormally large baby may make the pregnant female too heavy and therefore unable to hunt or avoid predators; it may also cause problems to both mother and baby during birth. Conversely, a very small baby may not be strong enough to walk/run/keep pace with its parents, and might therefore be vulnerable to predators.

2 **Directional selection** – this tends to occur following some environmental change which causes selection pressure. Organisms with a particular extreme of phenotype may have a selective advantage. The giraffe is a classic example: probably following a drought only the tallest animals could reach

Fig 4
Different types of selection: the shaded areas in the top row of graphs show individuals with a selective advantage.

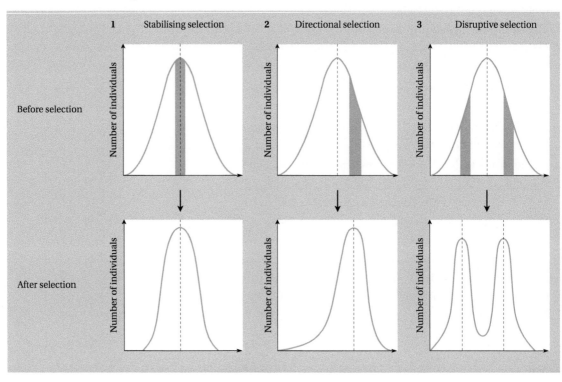

vegetation and so got enough food to survive and pass on their alleles/allele combinations to the next generation.

3 **Disruptive selection** – this is selection against the middle, favouring the individuals with extremes of phenotype. This was seen in certain species of finches on the Galapagos Islands. Starting with a common ancestor that showed a range in beak size, those with smaller beaks had an advantage when catching insects, while those with larger, stronger beaks had an advantage when breaking open seeds. Those with a phenotype on the mid range were out-competed by those at the extremes.

Speciation

Defining a species is difficult because new species evolve from existing species and the process is gradual. However, for examination purposes a working definition of a species is as follows:

> **Definition**
>
> *Species: a key definition*
>
> *A population or group of populations of similar individuals that can mate and produce fertile offspring.*

The classic example of this concept is the horse and the donkey. A male horse can mate with a female donkey to produce a mule. This hybrid, however, is sterile and so the horse and the donkey are regarded as separate species.

The evolution of new species

There are three key steps to the evolution of a new species:

1 **Isolation** – part of a population becomes isolated so that it *cannot breed* with the rest of the population.

2 **Natural selection** – acts in different ways according to the local situation to *change the frequency of alleles* and, eventually, the phenotypes.

3 **Speciation** – over the generations, genetic differences accumulate so that the different populations are unable to interbreed, even if brought together again.

Isolating mechanisms

We have seen that a key feature of speciation is isolation: where one population is prevented from interbreeding with another population. Isolating mechanisms are generally divided into two categories, although there are cases that don't fall neatly into either one:

- **Allopatric speciation** – the populations are physically separated and so cannot interbreed. Barriers include water, mountain ranges or, in recent times, man-made obstacles such as farms or even roads. This type of speciation is easy to explain and examples are common.

Notes

Domestic dogs come in all shapes and sizes, but a Jack Russell and a poodle can mate to produce perfectly fertile crossbreeds. We say that the poodle and the Jack Russell are different **breeds**, but not different species.

Fig 5
An example of speciation

Steps in the evolution of a new species:

1 A population of rabbits on an island is split by a new river, so the two populations are isolated.

2 Natural selection acts in different ways on the two populations: in the rocky area, for example, the rabbits can't burrow, so are more vulnerable to predators; natural selection therefore favours those with the keenest senses, best reflexes and greatest speed, and they evolve into hare-like creatures.

3 Over the generations, genetic differences accumulate so that even if the two populations come into contact, they are not able to breed successfully.

- **Sympatric speciation** – the populations are not physically separated, but still do not interbreed and evolve along different lines. Sympatric isolating mechanisms are nowhere near as clear–cut and easy to explain as allopatric isolating mechanisms. Examples are rare and arguments between scientists are common.

3.7.4 Populations in ecosystems

You will need to know the following important definitions.

Definitions

Populations and ecosystems: some key definitions

Ecosystem – *a natural unit consisting of producers, consumers and decomposers together with non-living components, for example, a pond, lake, coral reef or rainforest. The conditions within a particular ecosystem are usually fairly uniform.*

Population – *a group of individuals of the same species. The range of the population varies according to the species; the water fleas in a pond constitute a population, but so does the entire human population on Earth. Importantly, members of the same population can potentially interbreed.*

Community – *all of the organisms of all species in the ecosystem. The communities found in a particular **habitat** are based on dynamic feeding relationships. This means that the size of a population is determined by other populations that it preys on, or that prey on it.*

Habitat – *an organism's environment. For small organisms the immediate surroundings – the* **microhabitat** – *is often of vital importance. For instance, aphids (for example, greenfly) can usually be found on the underside of a leaf, next to a vein. If the individual moved just a millimetre away the conditions would change – there could be less food available, and the greenfly may be more exposed to wind movements.*

Niche – *a concept that explains an organism's place in the ecosystem. A niche is largely defined by what an organism eats (unless it is a plant), what eats it and what conditions it lives in. The* **competitive exclusion principle** *states that no two species can occupy precisely the same niche, so they don't compete for precisely the same resources.*

Ecosystems are unbelievably complex, and to even begin to understand what is going on we must take careful measurements, both of the organisms in an ecosystem and of the physical/chemical conditions that form their habitat. In this section, we look at some methods for **sampling** organisms and some techniques used to measure **abiotic factors** (physical, or non-living, features of the ecosystem).

Table 8
Methods of measuring some abiotic factors

Abiotic factor	Measured by	Effect of abiotic factor on organisms
Temperature	Thermometer/ thermal probe	Enzymes and therefore metabolism are temperature sensitive. Metabolic systems only work efficiently within relatively narrow temperature ranges. If the temperature is too low, metabolism slows; if it is too high, metabolism becomes imbalanced, and in extreme cases enzymes can be denatured.
Humidity	Hygrometer (a hand-held device)	Affects the rate of evaporation; the higher the humidity, the lower the rate of evaporation. The level of humidity determines the speed of evaporation, and so controls the effectiveness of transpiration in plants and thermoregulation (sweating/panting) in animals.
Light intensity	Light meter or light sensor	Often a limiting factor in photosynthesis – which affects the productivity of the whole ecosystem.
pH	pH meter or chemical (indicator) test	Enzymes are very pH sensitive – metabolism can be disrupted if conditions become too acidic or alkaline.
Oxygen concentration in water	Oxygen-sensitive electrode or chemical test	Oxygen is essential for aerobic respiration. Oxygen solubility in water is low and varies with temperature; the lower the temperature, the more oxygen will dissolve in it.
Carbon dioxide concentration	Gas analysis	Essential for photosynthesis – can be a limiting factor in some circumstances. Carbon dioxide is an acidic gas, so raised levels can affect aquatic ecosystems.
Wind speed	Anemometer (hand-held device)	Affects rate of evaporation and cooling, so has an impact on transpiration in plants and thermoregulation in animals.

Notes

In examination answers, throwing quadrats is not a satisfactory way of placing quadrats randomly (even if that's what you did on your field trip). You must say that you mapped out the area, created a grid and selected squares without bias by using a table of random numbers or a specific computer program.

There are two basic techniques commonly used in field studies: **quadrats** and **transects**. Quadrats allow you to sample different areas, while transects allow you to measure change from one area to another.

Quadrats

Frame quadrats are sample areas of ground that are small enough to be studied in a short time. A square frame of 50 cm × 50 cm is often used because the grid is manageable and portable (Fig 6a). Quadrats are normally used to compare one area of ground with another, such as the vegetation on north- and south-facing sides of a hill, or the species diversity on mown and unmown patches of ground. You obviously cannot count all the plants in the area, but by placing quadrats randomly you can sample a representative area. Methods of ensuring that the quadrats are random – i.e. without human

Fig 6

a Quadrats are usually used to sample different areas of ground; the area can be divided using a grid and quadrats are then placed in squares chosen at random. The experimenter may measure the occurrence (*Is the species present or not?*), number of individuals or the percentage cover of the different species. Percentage cover is most easily done using a quadrat that is divided into 100 smaller squares (perhaps 5 cm × 5 cm) and counting the number of squares in which the species is present.

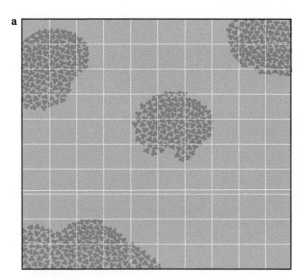

b Transects are lines that allow us to sample along a changing habitat. Different types of transects include the **interrupted belt transect**, where quadrats are placed at intervals along the line, a **continuous belt transect**, which is self-explanatory, or a **point transect** where just the species touching a particular point on the line are recorded.
NB: A point transect and a line transect are both transects that are done without the use of quadrats. However, with a line transect you can record every species that touches the line, while with a point transect you record the species that touches specific points (say every 25 or 50 cm).

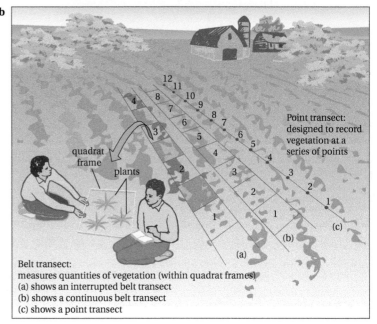

Point transect: designed to record vegetation at a series of points

quadrat frame

plants

Belt transect:
measures quantities of vegetation (within quadrat frames)
(a) shows an interrupted belt transect
(b) shows a continuous belt transect
(c) shows a point transect

bias – include mapping out the area into a grid pattern and selecting squares using random numbers from tables or from a computer program designed for this purpose.

Transects

Line transects and belt transects (see Fig 6b) are used to sample organisms along a line in order to show a change from one area to another, such as down a rocky shoreline or along sand dunes.

The most appropriate size of quadrat depends on the nature of the area being studied. A rainforest may require quadrats of 20 m square (achieved by pegging out rope/string), while lichens or mosses on a stone wall or tree trunk might only need a quadrat of 25 cm square.

Mark-release-recapture

Transects and quadrats are not a lot of use for estimating the population of animals that can move. It might work for limpets, but trying to throw a quadrat on a rabbit is a fruitless task. We need to be a little more devious.

1   Capture a number of individuals using a suitable trap (from those shown in Fig 7). Longworth traps are good for small mammals such as mice and voles. Pitfall traps may be more suitable for ground-dwelling insects.

2   Mark the individuals so that you will recognise them if you catch them again. Obviously, the animals should not be harmed, nor marked in such a way that they are less mobile or more visible to predators. (Marking snails with fluorescent dye may seem like a good idea at the time, but ...)

3   Release the marked animals.

4   Leave enough time for the marked individuals to redistribute themselves among the unmarked population, then capture a similar number and count the number of individuals that are marked and have therefore been caught before.

The population size can be estimated by the following formula, known as the **Lincoln index**:

$$\text{Population} = \frac{M \times C}{R}$$

Where:

M = total number of animals captured and marked on the first visit

C = total number of animals captured on the second visit

R = number of animals captured on the first visit that were then recaptured on the second visit (i.e. number in second sample that were marked).

Notes

Make sure you know the difference between **qualitative data** (in this context: what species are present) and **quantitative data** (how many individuals are present, or the percentage cover/relative abundance).

Remember that when collecting data, random sampling is important because it results in data which are unbiased and therefore suitable for statistical analysis.

Essential Notes

The **mark-release-recapture** method assumes that the population is stable, and that no births/deaths or emigration/immigration took place between the first and second samples. The calculation shown in the worked example is probably the simplest one that can be done. Repeated measurement over time, especially when animals are tagged/ringed and given individual numbers, can give valuable information about population size, long- and short-term change, migration patterns, etc.

Worked Example

On one night 90 bank voles, *Clethrionomys glareolus*, were captured and marked by cutting their fur in a small area (so the dark fur underneath showed up). On the second night 85 were captured, of which 14 were marked.

$$\text{Population} = \frac{90 \times 85}{14} = 546 \text{ individuals (rounded to nearest whole number)}$$

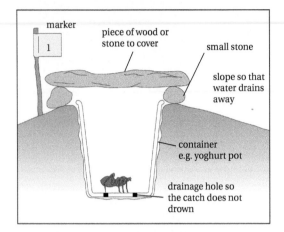

Pitfall traps can be used to trap invertebrates that are active on the soil surface or in leaf litter. 10% methanal (formalin) can be placed in the pitfall to kill predators that might otherwise kill other captives. Pitfalls are cheap and easy to use, but the number of individuals caught tends to reflect the activity of a particular species as well as its abundance.

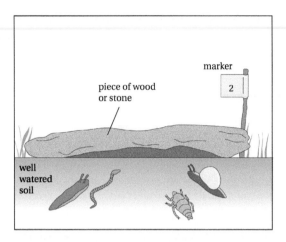

Cover traps are left for a few days before inspection. Like pitfall traps they can be baited with meat, jam or potato. The catch includes slow-moving animals like slugs, earthworms, snails and woodlice.

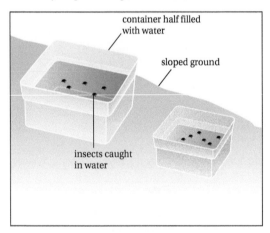

Water traps are left on open ground at different heights. Yellow traps seem to attract aphids while white attracts flies. They can be made from old ice cream cartons, half filled with water. Some washing up liquid can be added to reduce the surface tension so that insects landing on the water will sink.

Source: Adapted from Wiltshire Wildlife Trust

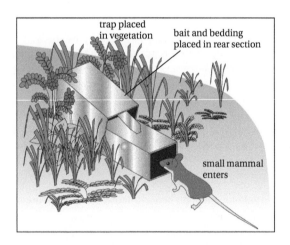

Longworth traps are baited with cheese or other food, and left in long grass where small mammals are likely to be found. As the animal enters, it triggers a lever which allows the trap door to close behind it. The trap should contain warm, dry bedding so that the animal comes to no harm before the trap is revisited and the animal is released.

Fig 7

Different types of traps. Ethical issues here include the fact that water traps kill most of the organisms that fall into them, and that shrews in Longworth traps may starve to death in a very short time.

To put this into perspective, consider the two extremes. If you capture 100 on the first night, and 100 on the second night, of which just one is marked, the population is about 10 000. (If none is marked, the formula will not work and the population is too large to estimate.) However, if you capture 100 on the second night that are all marked, then you have probably recaptured the entire population, which must be 100. Of course, this is unlikely.

Risk management

Any field work must be conducted safely, which usually involves some basic precautions. A risk assessment is needed before field work is undertaken.

Higher risk activities include:

- working near water, in remote countryside, wintery conditions, on or near cliffs or steep terrain
- working in an area where extremes of weather or sudden environmental change can occur
- overseas field work, which of course presents a greater risk in terms of more dangerous wildlife, exposure to more potentially fatal diseases and the availability of health care.

Water presents particular problems. Areas of risk include:

- slippery and sharp rocky shore work, compounded if the sea is rough because students may get swept off the shore by an abnormally large wave
- fast-flowing streams and rivers where students can be swept off their feet
- deep water where students could drown
- ponds or streams with steep and/or slippery banks
- polluted water – even a relatively clean looking stream or pond can contain a variety of dangerous bacteria.

Reasonable precautions

- Only work in places where the risks of falling in and of pollution are small and where it is shallow enough so that anyone falling in is unlikely to swallow water.
- Ensure that cuts or other forms of broken skin are covered by a plaster or gloves before the field work begins.
- Ensure that all participants – students and adults – wash their hands immediately afterwards. Packed lunches should be consumed before starting field work in fresh water.

Ethical issues

The major ethical issue with field work is that the ecosystem being studied should not be damaged. Large numbers of people all studying the same habitat can leave the populations of some species permanently damaged. For example, a whole year group of students performing a kick sample in a river or stream can seriously disrupt the population of many invertebrates. This in turn disrupts the whole food chain.

Overall:

- Students are expected to be aware of how they can cause minimum impact to the habitat and the organisms they are studying.
- There are opportunities to learn about and debate the conflict between human interest and the environment, including the pros and cons of habitat management and the importance of environmental sustainability.

Variations in population size

Suppose you introduce a pair of rabbits onto an island. How would the population grow? Assuming that they were a healthy pair, and they managed to reproduce, the population growth would be like that shown in Fig 8. This classic pattern is seen in many different situations; rats in a warehouse, beetles in a

Essential Notes

Weil's disease is caused by bacteria transmitted through rat's urine. Water polluted by farmyard waste, sewage and similar forms of effluent may be infected. Weil's disease can be very dangerous but the risk of contracting it from doing field work in fresh water is negligible provided that appropriate precautions are taken.

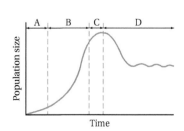

Fig 8
A generalised population growth curve

sack of flour, elephants in a game reserve, bacteria in your navel or humans on the planet; only the timescale changes. The size of a population depends on a combination of **biotic** and abiotic environmental factors.

Species diversity

The numbers of different species that form the community of an ecosystem can vary greatly. Environments such as a coral reef or a rainforest show a high **biodiversity** because conditions are generally favourable and stable. In these situations biotic factors – those due to other organisms – dominate organisms' lives. In contrast, in harsh environments such as the arctic or desert regions, there is a low biodiversity and abiotic factors, such as temperature and water availability, dominate.

The stages in the population growth curve (Fig 8) are:

A **Lag phase** – a time of slow growth. There are many different reasons for the lag; microorganisms may need time to activate genes and synthesise the enzymes needed to utilise a new food source, and in more complex organisms, species that reproduce sexually may take a while to grow and reach sexual maturity.

B **Log (logarithmic)** or **exponential phase** – a period of rapid and unrestricted growth, when conditions are favourable, for example, when there is plenty of food. The key point is that there are no **limiting factors**.

C Growth slows due to limiting factors. No population can go on increasing indefinitely; sooner or later there will be **environmental resistance** of some sort. Food may become scarce, waste may accumulate, etc.

D The population stabilises at its **carrying capacity**; the size of population that can be supported in a given area, for example, a pond may be able to support a population of 2000 water fleas, but not more.

Factors affecting populations

No matter what the species, no population can go on expanding indefinitely. The human population has been rising for thousands of years but even we cannot go on in the same way. If we carried on reproducing to our fullest capacity, a point would be rapidly reached when there was not enough food, too much overcrowding and disease. At the moment there are different rates of population growth in different parts of the world for various social and economic reasons.

Most populations are limited by a simpler set of factors than those that affect humans:

● **Competition** – all organisms are locked into a struggle to eat and not be eaten until they have a chance to reproduce. More organisms are born than can possibly survive and an inevitable consequence of this is competition. Common examples of competition include plants competing for light, water and minerals from the soil, while animals may compete for food, mates or nesting sites. There are two types of competition:

– **Interspecific** competition occurs in individuals from *different* species. Badgers and foxes, for example, may compete for some of the same food sources and burrows. Some plants secrete chemicals that inhibit the growth of competing species.

– **Intraspecific** competition occurs in individuals of the *same* species.

Notes

In questions about limiting factors, think carefully about the species involved. Organisms seldom run out of space. Food and water supplies tend to run out long before organisms are standing shoulder to shoulder.

Notes

To help you remember which type of competition is which: interspecific competition is **between** species like international football matches are **between** nations.

- **Predation** – this is also a key biotic factor. The populations of predator and prey are often closely linked (Fig 9), especially if the predator preys on one particular species, as in the case of ladybirds and aphids.

Essential Notes

Predator–prey relationships such as the one in Fig 9 are only this clear when the predator relies on one particular prey. Most relationships are more complex than this, as predators have more than one prey animal.

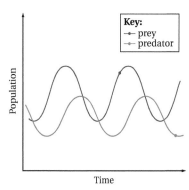

Fig 9

A simplified graph to show the relationship between predator and prey. When the prey population is high, there is a lot of food for the predators, whose population rises after a time lag. The large predator population takes more of the prey animals, whose population falls; consequently, there is less food for the predators, whose population then falls.

The development of an ecosystem

Before deforestation in the last thousand years or so, the UK was largely covered with **deciduous forest**: broad-leafed species that shed their leaves in the autumn. The dominant species were oak, ash, beech and birch, among others. This is the **climax community** that develops in our temperate climate. Bears, lynxes and wolves roamed the countryside; the human population was small and its influence was negligible. Now there is very little 'natural' forest left – virtually none in England – the forests that do exist were planted by humans.

So how did the forest develop in the first place? Ecosystems develop by the processes of **colonisation** and **succession** until the climax community is established (Fig 10).

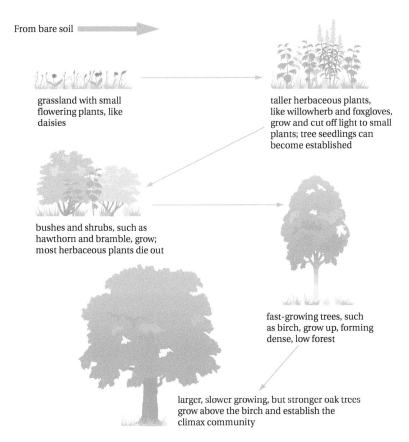

From bare soil

grassland with small flowering plants, like daisies

taller herbaceous plants, like willowherb and foxgloves, grow and cut off light to small plants; tree seedlings can become established

bushes and shrubs, such as hawthorn and bramble, grow; most herbaceous plants die out

fast-growing trees, such as birch, grow up, forming dense, low forest

larger, slower growing, but stronger oak trees grow above the birch and establish the climax community

Fig 10

Succession that occurs on a patch of bare ground, assuming little or no grazing. Deciduous (or broadleaf) trees lose their leaves in winter; in contrast some trees are evergreen, for example, the pine trees that dominate the forests of colder latitudes; herbaceous plants have no woody tissue, in contrast to shrubs and trees

Notes

In examination questions, avoid being anthropomorphic, i.e. giving human emotions to non-human organisms; for example, trees do not 'like' light, woodlice are not 'happy' in rotting vegetation and plants don't 'know' when it's time to flower.

Notes

It is a good idea to learn one example of succession in detail, and make sure that you can name some plant species at each stage, rather than talking vaguely about 'trees and shrubs', etc.

Essential Notes

The vital point when explaining how ecosystems develop is that existing species *make conditions more favourable*, allowing new species to become established and *out-compete* the existing ones.

The process of succession normally takes place over decades or centuries, but there are two common ways of studying the process:

- Clear a patch of ground and watch what happens to the bare soil.
- Study a sand dune system near a beach and observe the changes that occur as you move inland.

1. Clearing a patch of ground

Remove all the plants, fence it off from grazing animals, and observe the changes. The problem is that it takes at least 50 years, but eventually the forest is re-established. There are two conditions that need to be met if the ecosystem is to develop as normal:

- The soil must initially have a relatively low humus content.
- There must be no grazing animals that can get onto the plot.

The main stages that you would observe as the ecosystem develops are:

- **Colonisation** – the bare soil is colonised by what we might think of as weeds – herbaceous plants that can grow in relatively nutrient-poor soil. They have a rapid life cycle, then die back and increase the humus content of the soil. Typical coloniser species are grasses, daisies, dandelions and clover.

- **Succession** – this occurs because the colonisers change the habitat. Once the colonisers have improved the quality of the soil, by dying and rotting, more species can grow. The **diversity** of herbaceous plants increases greatly as the abiotic environment becomes more favourable. Taller plants cut off the light and so out-compete the shorter plants. The greater diversity of plants attracts more insects, which attract birds and small mammals. At this stage grazing can have a marked effect and prevent any further succession. Many plants, including tree saplings, have their growing points (meristems) at the top of the stem, and herbivores such as rabbits and sheep prevent any further growth. In contrast, grasses grow from the base of the stem, so they thrive despite constant grazing.

- **Establishment of the climax community** – in the next phase, small woody plants – shrubs such as hawthorn and bramble – begin to dominate. In turn, these are out-competed for light by fast-growing tree species such as birch that form a low, dense forest. Eventually the large, but slow-growing, trees – notably the oak – begin to dominate until the climax community is established and there is no further succession.

The climax community that develops depends on the climate. The process outlined above will not happen, for example, on exposed hillsides where the soil is too thin or the wind too harsh. Other examples of climax communities around the world include rainforest, cloud forest, tundra, grassland and coniferous forest.

2. Studying a sand dune system

Sand dunes are useful areas to study because you can observe colonisation and succession without having to wait 50 years. You can see some of the changes associated with the development of ecosystems as you simply walk inland (Fig 11). Near the sea the dunes are at their youngest – wind tends to pile up the sand and the profile changes from year to year.

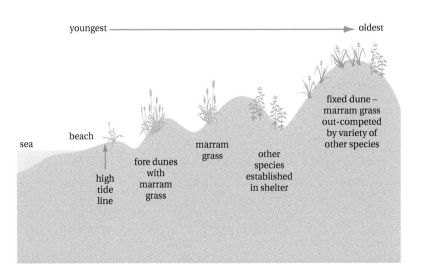

youngest ———————————————————→ oldest

sea
beach
high tide line
fore dunes with marram grass
marram grass
other species established in shelter
fixed dune – marram grass out-competed by variety of other species

Fig 11
Sand dune succession

Sand is a very difficult medium for plants: water and nutrients drain straight through, and the constant shifting makes it very difficult for the roots to anchor the plant. However some **pioneer species**, notably sand couch grass, lyme grass and marram grass in the UK, have a dense root system that binds the sand together. This holds water and humus particles and makes the whole dune more permanent. Sometimes the marram grass is the pioneer species, and sometimes the sand couch grass and lyme grass are the pioneers due to their greater tolerance to salt. In most dunes, marram grass is the key species because it makes the environment less hostile, so that other plants such as ragwort, willow and grasses can take over. As you move inland the sand gets darker because the humus content increases and so does the species diversity.

3. Succession from bare rock
Although our planet seems pretty crowded, there are still a few places where ecosystems can develop from scratch. A good example is the areas created by volcanic lava. As the lava cools it creates a barren, hostile environment but even this can be colonised by organisms such as bacteria, fungi, algae and lichens. (You can often observe the same process beginning on the tiles on your roof.) As the rock weathers, soil particles gather in cracks so that mosses and other small, shallow-rooted plants can get a hold. As these plants spread they create yet more soil, which increases in both depth and nutrient content. Succession then takes place in the manner described above.

Managing succession
An understanding of how to control ecological succession is important in a variety of situations:

- When growing trees and other plants in a **sustainable** way. Different trees grow at different rates, and appear at different stages of a succession. By harvesting fast-growing trees at the correct size and age, foresters can maximise sustainable wood production. This approach can also maximise biodiversity, which is often at its greatest during a succession rather than at the end. This increases the number of potentially useful plants and provides more habitats for animals. The basic stages of a succession were shown in

Essential Notes

Lichens consist of algae growing inside a fungus. The fungus provides anchorage and protection from drying out. In turn, the algae can photosynthesise and give the fungus organic compounds that would otherwise be unavailable from bare rock.

Fig 10. It can be seen that it will take a lot longer to harvest slow-growing trees such as oak, than it will to harvest the faster-growing trees such as birch.

- In some cases, succession is controlled because of its aesthetic value. For example, the appearance of the Lake District with its rolling landscapes is maintained largely by sheep farming – because grazing prevents succession back to woodland. Sheep farming is usually unprofitable but tourism is not, so farmers are actually paid subsidies to maintain the appearance of the countryside.

- Ponds are not usually permanent features. Over time, they 'silt up' with dead leaves and soil, etc., so that they dry out. Then the usual land succession takes over. However, human activity often reduces the number of new ponds that can develop, so old ponds can be maintained by dredging, thus preventing them from silting up. In this way, a valuable habitat is saved and biodiversity is maintained.

3.8 The control of gene expression

3.8.1 Alteration of the sequence of bases in DNA can alter the structure of proteins

Essential Notes

In exam answers, remember to write nucleotides and not bases. It is important to appreciate that, although the vital point with mutation is that the base sequence is changed, this happens because different **nucleotides** are added. All bases come with a sugar and a phosphate attached.

Gene **mutations** occur when DNA is damaged and not repaired, or is copied incorrectly due to errors in DNA replication. As a result of a mutation, the sequence of **bases** in the DNA is changed.

You need to know about six different types of gene mutation. You may remember the first three from Year 1 (see *Collins Student Support Materials AS/A-Level year 1 – Topics 3 and 4*):

Addition: one or more extra nucleotides are inserted, so all the other bases in one direction are pushed along. This is called a **frame shift**.

Deletion: one or more nucleotides are removed, so all the other nucleotides in one direction are effectively moved back. This also results in a frame shift.

Substitution: a nucleotide is replaced by a nucleotide with a different base. This changes just one **codon** and so does not result in a frame shift.

Inversion: Where a sequence of nucleotides is inserted backwards. The inversion can affect as little as two nucleotides or as much as whole genes or a non-coding sequence. Inversions occur when there are two breaks in a stretch of DNA and the repair mechanism rotates the section of DNA between the two breaks by 180 degrees before putting it back in. So, for this section of DNA, the sense and non-sense strands have effectively been swapped over. The usual result is a non-functional protein.

Duplication: where a sequence of nucleotides, or a whole gene or part of a gene, is added more than once. This is an important source of variation, because the repeated genes can mutate, while the original continues to make the required protein, or vice versa.

Translocation: where a base sequence is removed and inserted at a different place either on the same, or a different chromosome, so that part of a gene becomes attached to another gene. The new genes will normally make non-functional proteins. If this happens to **proto-oncogenes**, **tumours** may result (see page xx, The causes of cancer).

Examples of each type of mutation using a sentence with three-letter words to represent the base **triplets**:

Original sentence: THE OLD MEN SAW THE LAD

Substitution: THE OLD HEN SAW THE LAD

Deletion: THE LDM ENS AWT HEL AD

Translocation THE OLD SAW THE LAD MEN

Inversion THE OLD MEN WAS THE LAD

Duplication THE OLD OLD MEN SAW THE LAD

Addition: THE COL DME NSA WTH ELA D

3.8.2 Gene expression is controlled by a number of features

3.8.2.1 Most of a cell's DNA is not translated

Multicellular organisms such as humans start out life as one fertilised egg – a **zygote**. Looking at the **genome** of that cell, it is apparent that the genes only form a small amount of the total DNA. Most of the DNA is found between the genes and is known as *non-coding DNA*. This is never **transcribed** or **translated**. It has been estimated that between 1%–2% of the human genome actually codes for proteins. The rest has been described as junk DNA but it might have an important function in the expression of the coding sequences.

How does a single-celled embryo grow and develop into a complex, organised individual such as a human? We know that the secret lies in the control of *cell division* and *cell differentiation*. Cells know when to divide and when to remain in interphase. We also know that cells differentiate, or specialise, by the selective activation of certain genes. Cells that have the potential to differentiate into different specialised cells are called **stem cells**.

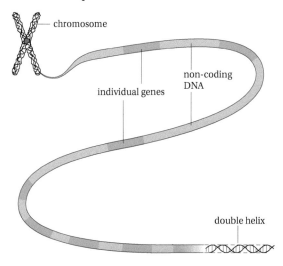

chromosome

individual genes

non-coding DNA

double helix

Fig 12
There is a lot of non-coding DNA between the genes, and some – the **introns** – are found within genes. Generally, the more complex the organism, the greater the percentage of non-coding DNA in the genome.

Different types of stem cells

Stem cells have two key properties:

1 **Self-renewal** – the ability to go through numerous cycles of mitosis while remaining undifferentiated.

2 **Potency** – the capacity to differentiate into specialised cell types.

Stem cells can be classed according to their potency – the range of cell types that they can produce. There is a sliding scale from the most potent to the least.

1 **Totipotent cells** can produce whole organisms. They can mature into any type of body cell, and the cell types associated with embryo support and development, such as the placenta. The zygote (a fertilised egg) is totipotent, and so are the cells in the early embryo.

2 **Pluripotent cells** – After about five days, the totipotent cells in the embryo begin to differentiate, or specialise, and form a hollow ball of cells called a **blastocyst**. The blastocyst has an outer layer of cells that eventually forms the placenta, and a cluster of cells inside the hollow sphere called the **inner cell mass**. The cells of the inner cell mass are pluripotent, meaning that they each have the potential to create every type of body cell, but not cells of the placenta. Pluripotent cells are not capable of making a whole organism.

3 **Multipotent cells** – Pluripotent cells soon undergo further specialisation into multipotent cells, which are usually referred to as **adult stem cells**. These cells can give rise to a limited number of other, specific types of cells. For example, haematopoietic cells (blood cells) in the bone marrow are multipotent and give rise to the various types of blood cells, including red cells, white cells and platelets.

4 **Unipotent cells** – can differentiate into just one type of specialised cell. Generally, they provide replacement cells for one type of tissue, such as gut lining or liver. One particularly promising line of research is the use of unipotent cardiac progenitor cells (CPCs) to make new cardiac muscle cells (**cardiomyocytes**). This would give us the potential to repair heart muscle damage after a myocardial infarction (heart attack).

Induced pluripotent stem cells (iPS)

Normally, differentiation of stem cells is a one-way process; once a cell has specialised, there is no going back. However, **induced pluripotent stem cells** (iPS cells) are produced artificially by 'reprogramming' unipotent cells to go back to the way they were in the embryo.

The process of turning specialised adult cells back into pluripotent stem cells, ones that can make *any* type of cell in the body, involves growing the cells with four specific **transcription factors**. In theory, any dividing cell of the body can be turned into a pluripotent stem cell.

These iPS cells have two big advantages over *embryonic* stem cells. First, they don't require the destruction of an embryo, which is an ethical dilemma for some. Second, iPS cells can be made from the patient's own cells, so are not likely to be rejected by their immune system.

Essential Notes

In the process of becoming specialised, stem cells translate only the relevant parts of their DNA. For example, if the genes that control liver formation are activated, the stem cell becomes a liver cell. But, if the genes that control muscle formation are turned on, that same totipotent cell would become a muscle cell.

Use of stem cells in treating human disorders

There are two well established uses of stem cells: making new skin for burns victims and bone marrow transplants for people undergoing cancer treatment. Both treatments involve the patient's own cells 'doing what they do', so there is no need for reprogramming and there's a reduced risk of rejection.

Other uses of stem cells are more experimental, and involve treating stem cells so that they will differentiate into one or more specific cell types that will reliably function in the body. As well as technical issues, there are a variety of other problems with these uses of stem cells including reliability, cost, ethics and legal issues. Stem cell therapy holds promise in the treatment of diabetes, for example, but scientists have yet to overcome the basic problem of persuading stem cells to differentiate into pancreatic beta cells. Trials are underway that show great potential in the treatment of various conditions including some types of blindness, heart disease, Parkinson's disease and nerve damage.

3.8.2.2 Regulation of transcription and translation

How are genes expressed?

Cells specialise by the selective activation of genes. In order for a gene to be **expressed**, the following steps have to occur:

- The gene is **transcribed** to make **messenger RNA** (**mRNA**).

- The **introns** are spliced out of the pre-mRNA (in eukaryotes).

- The code on the mRNA is **translated** into a protein.

- The protein may be modified to make it active, often in the Golgi body.

How is a gene activated?

Gene expression in eukaryotes is controlled by a variety of mechanisms that range from those that prevent transcription to those that act after the protein has been produced. The various mechanisms can be placed into four categories (Fig 13):

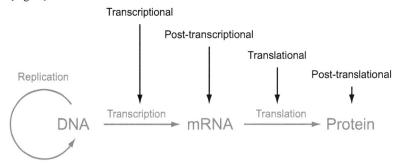

Fig 13
Regulation of gene expression

Transcriptional – these mechanisms prevent or promote transcription, and thereby turn off or turn on the synthesis of RNA.

Post-transcriptional – these mechanisms control or regulate mRNA after it has been produced.

Translational – these mechanisms prevent translation; they often involve protein factors that are needed for translation.

Post-translational – these mechanisms act after the protein has been produced.

Essential Notes

When transcription factors were first discovered, it was thought that they all stimulated transcription. Subsequent research has shown that factors that inhibit transcription are just as important in gene regulation. Transcription can be prevented by the presence of certain repressor molecules which attach to the promotor region and prevent the formation of the TIC.

The control of transcription

A gene will not be transcribed unless all of the required **transcription factors** are in place. Generally, most of the necessary factors are already present in the cell, and all that is needed is the addition of a particular molecule from outside the cell to 'complete the set'. Often, this molecule is a hormone.

Hundreds of different transcription factors have been discovered. Transcription factors are proteins that have DNA-binding domains that give them the ability to bind to specific sequences of DNA close to the gene, called **promoter regions** (Fig 14). Proteins have specific tertiary and quaternary structures, so transcription factors will have complementary shapes to specific exposed promoter region base sequences, or to other transcription factors that have already attached to the DNA.

Transcription begins when all of the transcription factors are assembled together, forming a **Transcription Initiation Complex (TIC).**

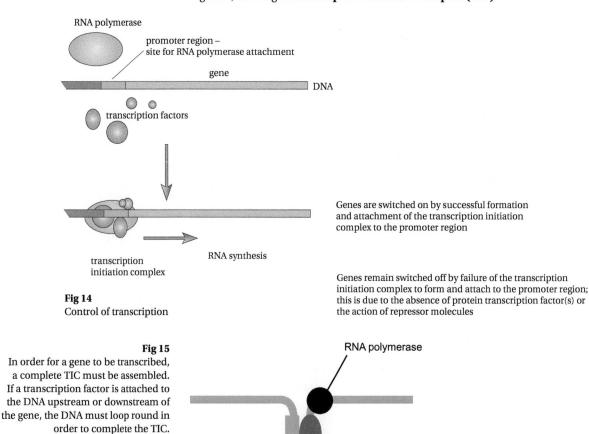

Genes are switched on by successful formation and attachment of the transcription initiation complex to the promoter region

Genes remain switched off by failure of the transcription initiation complex to form and attach to the promoter region; this is due to the absence of protein transcription factor(s) or the action of repressor molecules

Fig 14
Control of transcription

Fig 15
In order for a gene to be transcribed, a complete TIC must be assembled. If a transcription factor is attached to the DNA upstream or downstream of the gene, the DNA must loop round in order to complete the TIC.

Often, one of the vital factors is attached to the DNA in a position that is quite a distance away from the actual gene. In this case, the DNA must loop round in order to add the required factor to the TIC. This is shown in Fig 15. Once the last piece of the jigsaw is in place, and the TIC is complete, the **RNA polymerase** enzyme can race along the gene, transcribing as it goes (see *Collins Student Support Materials AS/A-Level year 1 – Topics 3 and 4*).

The role of oestrogen in gene transcription

Oestrogen is a steroid, a lipid hormone that passes through the cell surface membrane, through the cytoplasm and into the nucleus. Oestrogen binds with and activates oestrogen receptors (known as ERs – from the American spelling *estrogen*) in the nucleus of target cells.

The main function of these intracellular oestrogen receptors is as transcription factors which bind to DNA, thus regulating gene expression and stimulating the cell to make specific proteins.

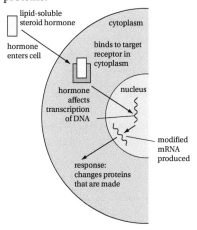

Fig 15
Steroid hormones bind to a receptor in the cytoplasm; the hormone/receptor complex enters the nucleus and binds directly to the DNA, altering gene expression; genes can be activated or de-activated, thereby changing the proteins made by the cell.

RNA interference

RNA interference (RNAi), sometimes called 'gene silencing', is a system that stops the expression of a particular gene by chopping up the mRNA before it can be translated (used to make a protein). There are several different classes of molecules that can bring about RNA interference. One method uses siRNA.

When foreign RNA (for example, viral RNA) gets into a cell, an enzyme called 'dicer' cleaves (splits) the foreign RNA, which is long and double-stranded (dsRNA), into short double-stranded RNA molecules that are about 21 nucleotides long. These molecules are called **short interfering RNA (siRNA)**. These siRNA strands then split into single-stranded RNAs (ssRNAs): the passenger strand and the guide strand. The passenger strand is not needed and is quickly broken down, but the guide strand becomes incorporated into an *RNA-induced silencing complex*, known as *RISC*.

RISC can be thought of as an RNA-destroying machine, but it does so selectively because of the guide strand attached. The guide strand allows RISC to seek out and bind only to those mRNA molecules that have complementary sequences.

RNAi is proving to be a very useful method of silencing harmful genes. Scientists have produced a whole range of RNAi molecules, which are very effective at silencing thousands of different genes.

Essential Notes

RNAi must be specific. It can't simply target and destroy all mRNAs because the cell wouldn't be able to make any of its vital proteins.

Essential Notes

Epigenetics is defined as the effect of environmental factors on gene expression.

Essential Notes

Twin studies can reveal a lot about epigenetics. Identical twins, who have identical genomes at birth, have been found to have different degrees of methylation and acetylation later in life. This is thought to be part of the reason that identical twins become more distinguishable as they get older.

Essential Notes

Epigenetics has been likened to going along the genome with highlighter pens. Red for most important and blue for less important.

Epigenetics

Epigenetics is a heritable change in gene function without changes to the base sequence of DNA. This basically means that the environment an individual is exposed to during its lifetime can cause genes to become labelled so that the likelihood of them being transcribed (and therefore expressed) increases or decreases. These changes can be passed on to its offspring.

There is no change in the sequence of DNA bases (that is, there are no mutations), but what changes is the potential for genes to be expressed.

There are two main ways in which transcription can be inhibited, both of which can be influenced by environmental factors:

- increased methylation of the DNA
- decreased acetylation of histones.

Methylation is simply the addition of a methyl (CH_3) group. In eukaryote DNA, methylation is the addition of the CH_3 group to the base cytosine, forming 5-methylcytosine. The process is carried out by DNA methyltransferase (DNMT) enzymes.

Methylation is an important method of epigenetic regulation: increased methylation inhibits transcription and, conversely, decreased methylation increases transcription. It is thought that when a particular promoter region is methylated, the transcription factors cannot bind and so the gene cannot be transcribed.

Histones are the 'organising proteins' around which the incredibly long DNA molecules are wound. DNA that is wound around a histone cannot be transcribed, so controlling the degree of attraction between DNA and histones can control transcription. **Acetylation** is one of several different histone modifications that can control transcription. Specifically, acetylation is the addition of an acetyl group ($COCH_3$) to the amino acid lysine on the histone. The acetyl group neutralises the positive charge on the lysine residues, reducing the attraction between the DNA and the histone. As a result, the DNA tends to detach from the histone, thus allowing transcription. Conversely, therefore, decreased acetylation of histones will decrease transcription, because the DNA will be more tightly bound to the histones.

Methylation is particularly significant to cancer researchers because inappropriate methylation of tumour suppressor genes switches them off, leading to uncontrolled cell growth (see section 3.8.2.3).

3.8.2.3 Gene expression and cancer

Gene mutations and cancer

Usually, cells in the human body only divide when they should, in order to allow growth or the repair of tissues. When the mechanisms that control cell division (mitosis) break down, the result is the uncontrolled division of cells, resulting in a tumour (Fig 16).

There are two types of tumours:

- **Benign** – these tumours are enclosed in a capsule and grow in the centre, so they do not invade the surrounding tissues; they are not cancerous and are often easily removed by surgery.

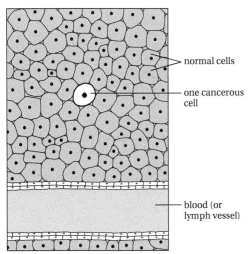

Fig 16
The development of a tumour

normal cells

one cancerous cell

rapidly dividing cancerous cells

blood (or lymph vessel)

One cell starts to divide uncontrollably: it becomes cancerous

The cancerous cell divides rapidly, forming a mass of cells – the primary tumour – which squashes out the neighbouring normal cells

- **Malignant** – these tumours grow at the edges, invading the surrounding tissues and organs; they are cancerous and are much more difficult to treat. Often it is difficult to tell where their boundaries are, making surgery difficult. Cells can break off and set up secondary tumours elsewhere in the body – this spreading process is called **metastasis**.

The causes of cancer

There are many factors involved in the development of cancer, including genetic and environmental factors.

Genetic factors
The development of cancer can be caused by a mutation of the genes that control cell division. Scientists have isolated several **proto-oncogenes**. When these mutate into **oncogenes** the cell loses its ability to control cell division.

The incidence of cancer, however, is much rarer than the mutation of oncogenes. This is because there is a back-up control system in the form of **tumour suppressor genes**. These genes prevent cells from dividing too quickly, giving time for the immune system to destroy the rogue cells, or for the damaged DNA to be repaired. If the tumour suppressor genes mutate, the cell's safety mechanisms are lost, and the development of cancer is more likely.

Cancer is more common in older people because their somatic (body) cells accumulate mutations. Sometimes, however, these mutations occur in gametes (sex cells), so that the genes are passed on to the next generation. People who inherit these genes are said to have a genetic predisposition to cancer; that is, they are more likely to develop cancer, especially at an early age.

Notes

Proto-oncogenes and tumour suppressor genes have been likened to the accelerator and brakes on a car.

The mutation of proto-oncogenes to oncogenes is like the accelerator being stuck down, and if both tumour suppressor genes mutate, it is as if the front and back brakes have both jammed. The combined effect is a loss of control.

Environmental factors

There are many factors that increase the rate of mutation of the cells described above, and so increase the risk of developing cancer. Such factors are called **carcinogens** and include:

- **Smoking** – tobacco smoke contains a variety of carcinogens.

- **Diet** – several substances we eat or drink can cause cancer, for example, alcohol has been linked with a higher incidence of mouth, throat and oesophageal cancer. A lack of fibre and a high intake of red meat and animal fat seems to be associated with cancers of the colon and rectum, which are very common in the Western world.

- **Oestrogen** – cells in some types of breast cancer have oestrogen receptors (see section 3.8.2.2). The binding of oestrogen with these receptors in cells in breast tissue initiates the transcription of genes for cell growth and division. This results in rapid division of the cells, and the formation of a tumour. This is why high oestrogen concentrations are linked with a higher risk of breast cancer.

- **Radiation** – certain types of radiation are known to be carcinogenic because they damage DNA. Ultraviolet radiation (in sunlight) does not penetrate far into human tissue but it can cause skin cancer. Ionising radiation, for example, from nuclear fallout, can penetrate much further and can cause cancers such as leukaemia (cancer of the bone marrow).

- **Chemical carcinogens** – asbestos, benzene, methanal (formaldehyde) and diesel exhaust fumes are all carcinogenic.

- **Microorganisms** – viruses, in particular, have been linked with the development of cancers. The *human papilloma virus* (*HPV*) is associated with over 90% of cases of cervical cancer.

3.8.3 Using genome projects

A **genome** is defined as 'the entirety of the genetic sequences in an organism'. In short, it's all of the DNA. The proteins that an organism can make from the genetic code are called the **proteome**. The human genome consists of 23 chromosomes, about 21 000 genes – it is still not certain how many – and three billion (3×10^9) base pairs. In humans, the entire genome is present twice in each cell.

It took 11 years to sequence the first human genome. However, progress in science is often limited by the technology available and nowhere is this truer than in genomics. There are now machines that can sequence an entire human genome in a few hours, and we are continually refining technologies that allow sequencing machines to work faster. As well as improved speed, the sequencing cost per base is becoming dramatically lower. When this technology is fully developed, it may be possible, and commonplace, to sequence an individual's entire genome in a matter of minutes.

One of the most far-reaching implications of the human genome project is the possibility of rapid and accurate genetic analysis of an individual's genotype, and the development of sensitive DNA probes to screen individuals for certain genetic diseases or personalised medicine to account for variation in how individuals respond to pharmaceuticals (see section 3.8.4.2).

3.8.4 Gene technologies allow the study and alteration of gene function, allowing a better understanding of organism function and the design of new industrial and medical processes

3.8.4.1 Recombinant DNA technology

Manipulating genes

This section is about what is generally referred to as genetic engineering. The genetic code is *universal* – the same codons code for the same amino acids – and proteins are made in much the same way in all organisms. This means that we can put a piece of DNA from one organism into another and be sure that it will still code for the same polypeptides and proteins.

Using genetic engineering, we can:

- find genes or specific DNA sequences
- cut them out
- make millions of copies of them in a very short time
- make artificial genes by working backwards from the protein or the mRNA
- insert genes into other cells and organisms so that they are expressed
- extract the product of the genes.

This section looks at each process in turn.

How do we find the genes we want?

How do you find a specific gene in an entire genome? One way is to make a **DNA probe**, which is a single-stranded piece of DNA with a label attached; the DNA base sequence is complementary to part of the gene that we want to find. When a DNA probe is mixed with single-stranded DNA from the genome, the DNA probe **hybridises** (joins) with the complementary strand of the section of DNA, showing its position. When the gene has been found, it can be cut out. (DNA probes are also used to identify short DNA sequences in medical diagnosis and in DNA fingerprinting, see sections 3.8.4.2 and 3.8.4.3.)

A problem with cutting eukaryote genes out of their genomes is that they still contain **introns**. Prokaryotic DNA does not contain introns, and therefore bacteria do not have the mechanisms for splicing them out. If a gene is inserted into bacteria, and transcribed and translated without splicing out the introns, the protein will be very different and almost certainly will not function. For this reason, it is much better to make artificial genes by working backwards from the mature mRNA or the protein.

To make an artificial gene from mature mRNA, we need to find cells that are actively expressing that gene. Human growth hormone, for example, is synthesised in the anterior lobe of the pituitary gland. The cytoplasm of these cells contains mRNA for growth hormone. This mRNA can be extracted and used to make a strand of **complementary DNA (cDNA)** using the enzyme **reverse transcriptase** (Fig 18).

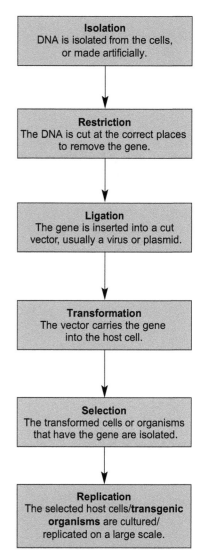

Isolation
DNA is isolated from the cells, or made artificially.

Restriction
The DNA is cut at the correct places to remove the gene.

Ligation
The gene is inserted into a cut vector, usually a virus or plasmid.

Transformation
The vector carries the gene into the host cell.

Selection
The transformed cells or organisms that have the gene are isolated.

Replication
The selected host cells/**transgenic organisms** are cultured/replicated on a large scale.

Fig 17 An overview of gene transfer

Fig 18
HIV. The human immunodeficiency virus contains RNA and the enzyme reverse transcriptase. Making DNA from RNA is transcription in reverse.

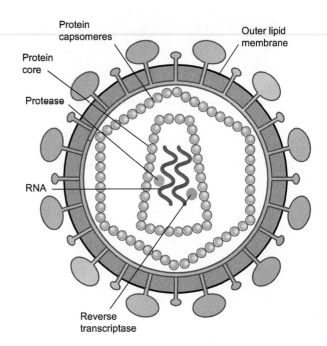

Essential Notes

A DNA probe can be radioactively labelled, and the label visualised using photographic film, or labelled with an enzyme that fluoresces (produces visible light) when a particular substrate is added.

The development of 'gene machines' has made it possible to make a piece of cDNA from the mRNA strand in a matter of hours. As an alternative to using mRNA, genes can be synthesised by working backwards from the primary sequence of protein. If you know the amino acid sequence of the protein, today's DNA synthesisers will make the gene that will code for that protein.

Fig 19
Producing cDNA from mRNA

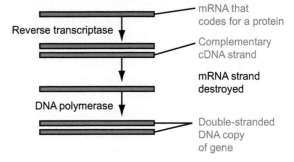

Cutting DNA out of the genome

To remove DNA from a cell, the cell membrane needs to be disrupted and the nucleus broken open. Once the DNA is free, the proteins that are associated with the DNA (histones) are removed using digestive enzymes.

Once the DNA has been isolated from the rest of the cell, the part of the DNA molecule that contains the required gene has to be cut out and the rest of the DNA discarded. This is important because genetic engineering must be precise; only known genes should be transferred from the donor organism.

The required gene is cut out by using **restriction endonuclease** enzymes, which cut DNA molecules at specific positions. Think of them as molecular scissors. Several different restriction enzymes occur naturally in bacteria. Their function is to chop up and destroy the DNA of any viruses that infect the bacterial cell.

Each enzyme cuts across the double-stranded DNA molecule at a specific nucleotide sequence, known as the **restriction site**. For example, one enzyme, known as EcoR1 cuts the strands only at the sequence shown in Fig 20.

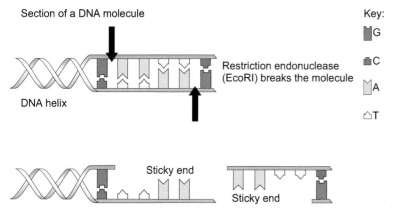

Section of a DNA molecule

Restriction endonuclease (EcoRI) breaks the molecule

DNA helix

Sticky end

Sticky end

Key:
G
C
A
T

Fig 20
The action of an endonuclease

Essential Notes

Restriction endonuclease enzymes are so called because they *restrict* viral growth by cutting *within nucleic acids*.

Restriction sites tend to be **palindromic**. In biology, this means that the base sequence in one direction reads the same on the other strand in the other direction. So if one side reads GTTAAC, the other strand will read CAATTG.

Most restriction enzymes do not cut straight across a DNA molecule – they separate the strands over a stretch of four bases, leaving each part of the broken DNA molecule with a short, single-stranded tail. These tails are called **sticky ends** (Fig 21). The advantage of sticky ends over clean cuts is that you have control over which DNA strands join – a sticky end can only join to a complementary sticky end.

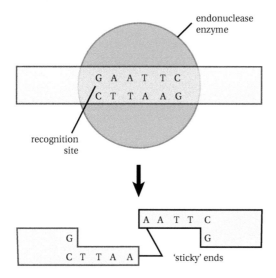

endonuclease enzyme

G A A T T C
C T T A A G

recognition site

A A T T C
G

G
C T T A A

'sticky' ends

Fig 21
Some restriction enzymes make staggered cuts, rather than clean ones, so that a few bases are exposed – four in this example; this allows the cut to be connected to any other piece of DNA that has been cut with the same enzyme. Note that the **recognition site** is palindromic.

Making multiple copies of the gene

Fragments of DNA can be amplified by *in vitro* and *in vivo* techniques. The *in vivo* technique involves putting the desired piece of DNA into a **vector** such as a **plasmid** or virus, and then using the vector to insert the DNA into a host cell such as a bacterium. The transformed host cell will (sometimes) adopt the new DNA and express it to make a particular protein (Fig 22). As each host cell divides in the culture medium, it produces genetically identical cells containing its own DNA and also the inserted or donor gene.

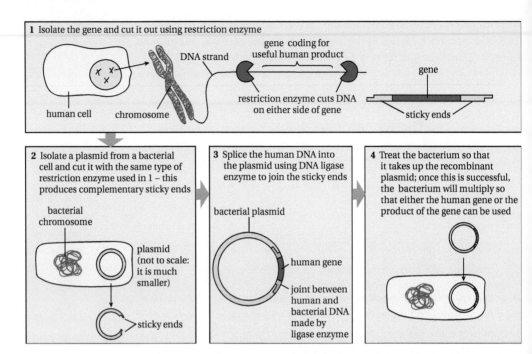

Fig 22
The process of *in vivo* cloning

One example of this is the manufacture of human insulin. The human insulin gene is isolated and multiple copies are made, using PCR. Then the gene must be inserted into a bacterium, which is achieved using plasmids as vectors.

The problem with this process is its unreliability. When plasmids are mixed with bacteria, for every bacterium that takes up a plasmid, tens of thousands do not. So how do you tell which bacteria have accepted the new gene? One way is to add a **marker gene** – an extra gene inserted into the plasmid along with the gene that is to be cloned.

A common example of a genetic marker is a gene for antibiotic resistance. The bacteria are grown on a medium that contains the antibiotic. Only those bacteria that took up the plasmid with the new gene and the gene for antibiotic resistance will survive and grow. Another example of a genetic marker is a gene which makes an enzyme that makes a coloured product. The bacteria are then grown with the required substrate in the growth medium, and any transgenic colonies will show up and so are easily seen and collected.

The polymerase chain reaction (PCR)

The **polymerase chain reaction (PCR)** is gene **cloning** *in vitro* (in a test tube). Copies make more copies so the process proceeds at an exponential rate: 1, 2, 4, 8, 16, and so on. PCR can amplify tiny amounts of DNA by a factor of over a million within an hour.

Uses of PCR include:

- Forensic analysis. PCR allows very small samples of DNA to be amplified (copied), and generate enough DNA for **genetic fingerprinting** to be carried out (see section 3.8.4.3).

Essential Notes

Plasmids are small circles of DNA that are found in bacterial cytoplasm.

Essential Notes

Marker genes are used in genetic research to identify which cells have incorporated the new DNA successfully. The marker genes remain in genetically modified (GM) organisms such as the commercially grown transgenic plants.

- Detection of genetic defects (see section 3.8.4.2). Using PCR means that only a few, or even just one cell from an embryo or a foetus is needed for the presence of particular alleles to be detected.

- Tissue typing prior to transplants.

- Analysis of tissue from extinct animals, such as the mammoth or the Tasmanian wolf, to establish evolutionary links.

- Paternity testing.

- Identification of human remains following a fire, or in cases of advanced decay, for example.

- Analysis of ancient human bodies in which some soft tissue has been preserved; these studies are helping us to understand the migrations of early human populations.

The process of PCR needs just four components:

1. The original DNA to be amplified.

2. Nucleotides.

3. A **DNA polymerase** enzyme. The enzyme **Taq polymerase** is commonly used. It comes from the bacterium *Thermus aquaticus*, which lives in hot volcanic springs. The enzyme is therefore thermostable and is not denatured by high temperatures.

4. Primers. These are short, single-stranded pieces of DNA that bind to the original DNA and signal to the enzyme where to start copying.

Essential Notes

Thermostable enzymes work at high temperatures without being denatured.

PCR amplifies DNA in a series of cycles. Each cycle takes about two minutes and consists of three stages (Fig 23).

1. **Denaturation**. The reaction mixture is heated to 94–98 °C for 20–30 seconds so that the hydrogen bonds break and the two strands of DNA separate (or 'melt') – so the DNA is single stranded.

2. **Annealing**. The temperature is reduced to 50–65 °C for 20–40 seconds so that the primers anneal (stick) to the single-stranded DNA.

3. **Extending**. The temperature is raised again to around 72 °C (which is within the optimum temperature range for Taq polymerase), and free nucleotides bind to complementary bases on DNA by specific base-pairing. This binding of nucleotides is orchestrated by the Taq polymerase, which moves along each separate single strand of DNA, adding complementary nucleotides and making double-stranded DNA.

So, at the end of the first cycle there will be two double-stranded copies for every original, then after the next cycle there will be four copies, then after the next cycle eight copies, and so on. The whole process is automated.

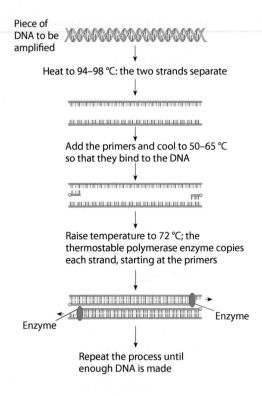

Fig 23
A single cycle of the polymerase chain reaction

Piece of DNA to be amplified

Heat to 94–98 °C: the two strands separate

Add the primers and cool to 50–65 °C so that they bind to the DNA

Raise temperature to 72 °C; the thermostable polymerase enzyme copies each strand, starting at the primers

Enzyme

Enzyme

Repeat the process until enough DNA is made

Table 8
The relative advantages of *in vivo* and *in vitro* gene cloning (amplification)

	In vivo **(in organism)**	In vitro **(using PCR)**
Advantages	Gene can be expressed (used to make protein); proofreading enzymes correct copying mistakes	Very quick; little purification of final sample needed
Disadvantages	Relatively slow, with more complex purification	Mistakes in copying the base sequence are common

Getting the gene into the host cell

So now we have lots of copies of our target gene. It is important to appreciate that if the gene is to be expressed, it also needs to have **promoter** and **terminator** regions to start and stop transcription. However, if the gene has been constructed using mRNA or protein sequencing methods, then it will only contain coded information for the primary structure of the protein. So, in order for it to be expressed correctly in the host, both promoter and terminator sequences need to be added to the gene sequence.

To get the genes into the host cell (a bacterial cell, for example), a vector is needed; a piece of DNA that can carry the gene into the host cell. Bacterial **plasmids** are commonly used. Plasmids are very useful in genetic engineering because these loops of DNA can replicate independently from the bacterial chromosome.

The plasmid is first cut open using the same restriction enzyme that was used to cut out the gene from the donor DNA. This creates a broken loop of DNA with sticky ends that are complementary to those of the donor gene. The donor gene can then be inserted into the plasmid loop using the enzyme DNA **ligase**, which catalyses the reaction that joins two sections of DNA (Fig 24).

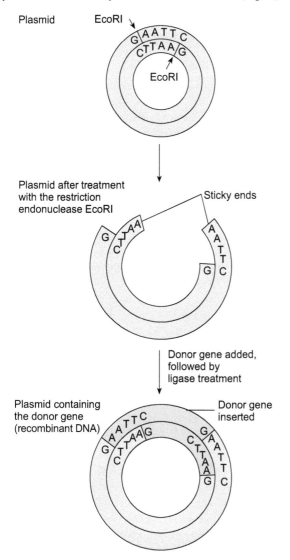

Fig 24
Inserting a gene into a plasmid

Getting the gene into the host cell DNA

The process of getting the new DNA incorporated into the genomes of living cells is called **transformation**. There are many different methods of transformation, and techniques are changing all the time. They are beyond the needs of the specification.

Gene therapy

So far we have looked at transferring genes into bacteria, but we can also transport genes into the cells of eukaryotes, including humans.

Gene therapy is one area of research that aims to treat genetic disease by giving patients healthy copies of defective alleles. However, achieving this is easier said than done. It is not too difficult to find and make lots of copies of the healthy allele – the techniques involved are outlined above – but the problem lies in getting the genes to the exact cells that need them, and making sure that they are expressed.

One example of a disease that might feasibly be treated using gene therapy is cystic fibrosis. Cystic fibrosis is caused by the effects of mutations in a gene that codes for a protein called CFTR. This molecule is one of the essential channel proteins in cell membranes, and it consists of 1480 amino acids. The function of the CFTR channel protein is to transport chloride ions through the cell membrane.

Gene therapy for cystic fibrosis involves inserting functional CFTR genes into the epithelial cells of the lungs, to replace the defective alleles in individuals who have two mutated alleles.

CFTR genes are made by one of the techniques outlined above ('gene machines'), cloned by PCR and either:

- inserted into plasmids, which are then inserted into liposomes – tiny spheres of lipid that can easily be absorbed through cell surface membranes
- inserted into harmless viruses that transfer genes into cells.

A major problem with gene therapy is getting the genes to the cells that need them. With cystic fibrosis, the task is relatively easy because the genes are needed in the cells lining the lungs, and so they can simply be inhaled. However, even when the genes do get to the cells, there is no guarantee that they will be expressed to make the missing protein.

3.8.4.2 Differences in DNA between individuals of the same species can be exploited for identification and diagnosis of heritable conditions

A **DNA probe** can be used to find a particular sequence of DNA. This probe is designed as a piece of single-stranded DNA that is complementary to the base sequence of the specified stretch of DNA.

Labelled DNA probes can be used to find specific alleles or base sequences. The probes are made from cDNA and will **hybridise** with (bind to) the target sequence and indicate its presence. This technique has great potential for identifying whether individuals are carrying an allele for an inherited condition, such as Huntington's disease, or genes that predispose an individual to an increased risk of certain diseases, such as prostate cancer or coronary heart disease.

An individual's genotype can also determine how they will respond to certain drugs. As a consequence of genome analysis, drugs and other medical treatments can be tailor-made to an individual's genetic make-up. Personalised medicine can, for example, allow a doctor to select a particular monoclonal antibody for a cancer patient (as described in section 3.2.4 in *Collins Student Support Materials AS/A-Level year 1 – Topics 3 and 4*). Although this approach is relatively new, it has the potential to make treatment a lot more reliable.

Generally, the process of **screening** refers to the process of testing all of the individuals in a certain cohort, to see whether they show symptoms of the disease or not. Currently, women of particular age groups in the UK are offered screening for cervical cancer and breast cancer. The development of more sophisticated DNA probes and genome analysis means that in the future we will feasibly be able to screen routinely for genetic conditions, even from birth. However, given the seriousness of some diagnoses, it is likely that screening will be linked with genetic counselling so that the individual is helped to deal with the possible consequences.

3.8.4.3 Genetic fingerprinting

The basic idea behind **genetic fingerprinting** (more accurately called DNA profiling) is that between the genes are long segments of non-coding DNA. Mutations and recombinations can accumulate in these regions. Specifically, non-coding regions contain many **variable number tandem repeats (VNTRs)**, which are particular base sequences repeated a certain number of times. Different people have different numbers of repeats, and so the length of the fragments between restriction sites will vary between individuals (Fig 25). More repeats means longer fragments between restriction sites.

The probability of two individuals having the same VNTRs is very low, although closely related people will share some sequences. More recently, DNA profiling has focused on *short tandem repeats (STRs)* because they are less likely to degrade over time and therefore give more reliable results with older DNA samples.

PCR can be used to amplify very tiny starting amounts of DNA from a sample (such as one hair root).

Essential Notes

Within VNTRs are core sequences that are common to all humans. It is the number of times the sequences are repeated that varies from person to person.

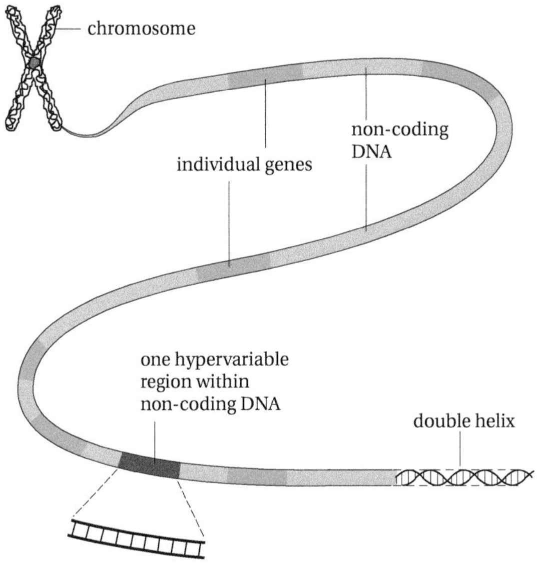

within a hypervariable region a core sequence is repeated a certain number of times: different people have different numbers of repeats and therefore different-sized hypervariable regions

Fig 25
Although each chromosome contains hundreds or thousands of genes, they only account for a small proportion of the length: over 95% is non-coding DNA; within these segments are hypervariable regions. These are unique to each individual and, when they are separated, the pattern they form provides the basis of DNA profiling.

Essential Notes

If nucleotides coding for fluorescent proteins are used in PCR, the resulting fragments are easy to show up after electrophoresis.

Essential Notes

Often, a DNA reference ladder is run in one lane of an electrophoresis gel. This contains DNA fragments of known size, and can be used to estimate the size of the unknown fragments.

Processing the DNA and separating the fragments

Once you have a DNA sample, the next stage is to cut the DNA into short lengths with a particular restriction enzyme (see section 3.8.4.1). The same restriction enzyme must be used for each DNA sample, so that it is certain that any differences in DNA fragments are due to differences in base sequences, rather than different enzymes cutting at different places.

The DNA fragments are then separated by **electrophoresis** (Fig 26). The mixture of DNA fragments is placed in a trough at one end of a long piece of agar gel that is placed in a container with a dilute solution of ionic salts. Electrodes are placed in the solution at either end and a potential difference is applied. The phosphate groups in the fragments of DNA give them a negative charge, so they are attracted through the gel towards the positive electrode. The smaller fragments move more rapidly through the gel matrix than the larger ones, so the different-sized fragments are separated in much the same way as they are in chromatography.

As shown in Fig 27, a **DNA probe** is then used to reveal the positions of the bands on the sheet containing the DNA sequence of interest. A general probe can be used to make all the bands show up. This allows us to see the familiar barcode pattern of DNA bands, which can be compared side by side with the DNA pattern of another sample.

Fig 26
Electrophoresis

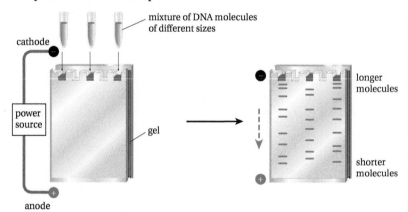

Fig 27
An overview of DNA profiling

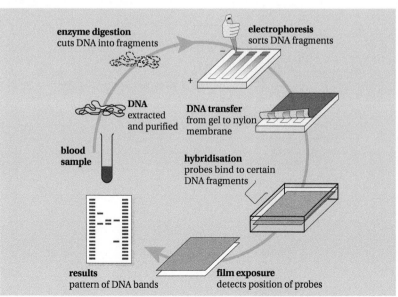

Uses of DNA profiling

As described in the PCR section above (3.8.4.1), analysing DNA fragments that have been cloned by PCR is used in forensic science, medical diagnosis for genetic defects, and to establish evolutionary relationships between living and extinct species. DNA profiling is also used in:

- Determining the genetic variability within a population. With both plants and animals, when a population is small, inbreeding becomes a danger. DNA profiling can be used to make sure that the most distantly related individuals breed together. Zoos keep stud books which contain this information so that small populations of endangered animals are kept as genetically diverse as possible.

- Paternity/maternity testing, and also to establish more distant links, which may be particularly important in establishing family trees, or in immigration cases (Fig 28).

- In a similar way to the point above, an animal's pedigree can be established. This can be very important when, for example, race horses are sold for large sums of money.

- Animal trafficking. For example, it is illegal to take many species of parrot from the wild, so it is only legal to sell parrots that have been bred in captivity. DNA analysis allows us to identify those taken from the wild.

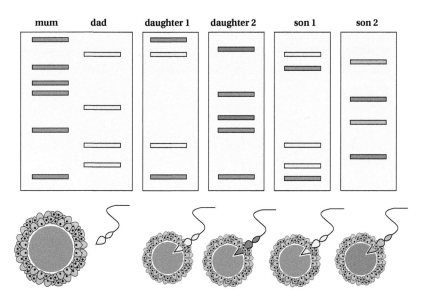

Fig 28
DNA profiles from one particular family. We inherit our DNA from our parents, and any bands that we do not get from our mother must come from our father. In this example, daughter 1 and son 1 clearly have their parents' DNA; daughter 2 is a step-daughter from mum's previous partner, while son 2 is adopted.

Practical and mathematical skills

In both the AS and A level papers at least 15% of marks will be allocated to the assessment of skills related to practical work. A minimum of 10% of the marks will be allocated to assessing mathematical skills at level 2 and above. These practical and mathematical skills are likely to overlap to some extent, for example applying mathematical concepts to analysing given data and in plotting and interpretation of graphs.

The required practical activities assessed at AS are:

- Investigation into the effect of a named variable on the rate of an enzyme-controlled reaction
- Preparation of stained squashes of cells from plant root tips; setup and use of an optical microscope to identify the stages of mitosis in these stained squashes and calculation of a mitotic index
- Production of a dilution series of a solute to produce a calibration curve with which to identify the water potential of plant tissue
- Investigation into the effect of a named variable on the permeability of cell-surface membranes
- Dissection of animal or plant gas exchange system or mass transport system or of organ within such a system
- Use of aseptic techniques to investigate the effect of antimicrobial substances on microbial growth.

The additional required practical activities assessed only at A level are:

- Use of chromatography to investigate the pigments isolated from leaves of different plants, for example leaves from shade-tolerant and shade-intolerant plants or leaves of different colours
- Investigation into the effect of a named factor on the rate of dehydrogenase activity in extracts of chloroplasts
- Investigation into the effect of a named variable on the rate of respiration of cultures of single-celled organisms
- Investigation into the effect of an environmental variable on the movement of an animal using either a choice chamber or a maze
- Production of a dilution series of a glucose solution and use of colorimetric techniques to produce a calibration curve with which to identify the concentration of glucose in an unknown 'urine' sample
- Investigation into the effect of a named environmental factor on the distribution of a given species.

Questions will assess the ability to understand in detail how to ensure that the use of instruments, equipment and techniques leads to results that are as accurate as possible. The list of apparatus and techniques is given in the specification.

Exam questions may require problem solving and application of scientific knowledge in practical contexts, including novel contexts.

Exam questions may also ask for critical comments on a given experimental method, conclusions from given observations or require the presentation of data in appropriate ways such as in tables or graphs. It will also be necessary to express numerical results to an appropriate precision with reference to uncertainties and errors, for example in thermometer readings.

The mathematical skills assessed are given in the specification.

Practice exam-style questions

1 A group of students set out to investigate the distribution of seaweeds and animals on a rocky shore. They carried out a continuous belt transect along an exposed and a sheltered rocky shore and mapped out their findings (Fig E1).

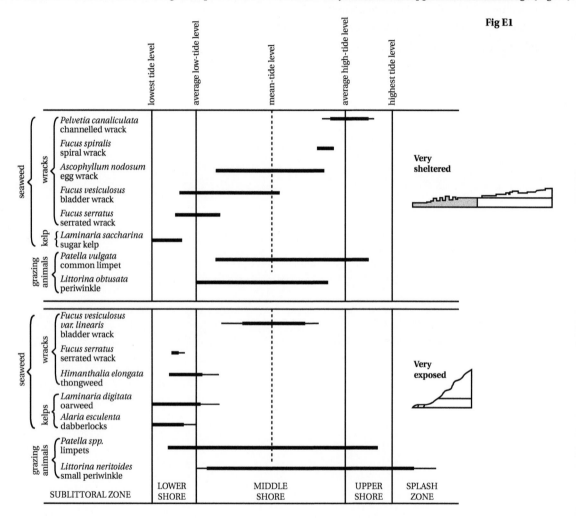

Fig E1

(a) Give one ethical and one safety consideration that would need to be taken into account before starting such an investigation.

_____

_____ 2 marks

(b) Explain how a continuous belt transect is carried out.

_____

_____

_____ 3 marks

(c) Explain why a belt transect was a more suitable choice of sampling method than throwing quadrats.

_____

_____ 2 marks

(d) Does the transect give qualitative or quantitative data? Explain your answer.

_____

_____ 2 marks

(e) **(i)** Name two species that are only found on the sheltered shore.

_____ 1 mark

(ii) Suggest two abiotic factors that could account for the distribution of these species.

_____

_____ 2 marks

Total marks: 12

2 A group of students investigated the effect of grazing by rabbits. They wanted to know if grazing had an effect on species diversity. They compared two patches of land – one that was grazed by rabbits and one that was fenced off.

(a) Suggest a null hypothesis for this investigation.

_____ 1 mark

(b) Explain how you would place the quadrats without bias.

_____

_____ 2 marks

Concentrating on the seven most abundant species in the area the students recorded the average number of individuals per quadrat, as shown in Table E1. They performed a statistical test to see if there was a significant difference in the distribution.

Table E1

Species	Grazed	Ungrazed	Significance (value of p)
Grass 1 (Agrostis)	25	21	0.5
Grass 2 (Festuca)	14	16	0.5
Buttercup	7	8	0.5
Dandelion	4	9	<0.05
Willow	0	11	<0.01
Bilberry	2	9	<0.01
Trefoil	0	4	<0.01

(c) What does the word 'significant' mean when applied to statistics?

_____ 1 mark

(d) Suggest which statistical test was used on these results.

_____ 1 mark

(e) What conclusions can be drawn from the results?

_____

_____

_____ 3 marks

(f) If the investigation were carried on for the next few decades, describe the changes that would occur in the fenced off (ungrazed) patch of land.

_____

_____

_____

_____

_____ 5 marks

Total marks: 13

3 *Paramecia* are single-celled protoctists found in ponds and other waterways. They feed on algae and other microscopic organisms. In a classic experiment, two species of *Paramecia* were grown both separately and together. The graphs (Fig E2) show the population growth curves for both species in both investigations.

Fig E2

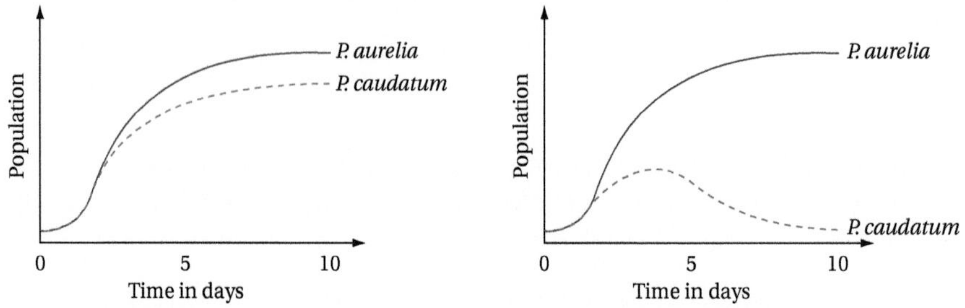

Graph 1 – grown separately

Graph 2 – grown together

(a) Suggest why growth is slow for the first day or so.

_____

_____ 2 marks

(b) Account for the rapid growth of both species between days 1 and 4 when grown separately.

_____

_____

_____ 3 marks

(c) Suggest an explanation for the pattern of population growth in Graph 2.

_____

_____ 2 marks

Total marks: 7

4 The graph (Fig E3) shows the population sizes of the snowshoe hare and the Canadian lynx over an 80-year period, taken from the records of Canadian fur trappers.

(a) What year saw the biggest fall in snowshoe hare numbers?

_____ 1 mark

(b) Describe and explain the relationship between the two populations.

_____

_____

_____

_____ 4 marks

Fig E3

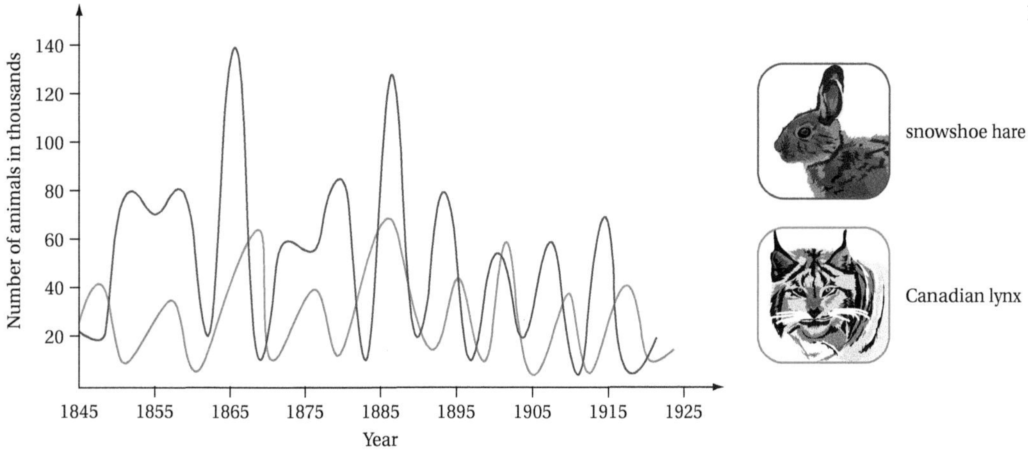

snowshoe hare

Canadian lynx

(c) Explain why relationships between predator and prey are rarely as clear as the one illustrated above.

_____

_____ 2 marks

Total marks: 7

5 The diagram below (Fig E4) shows a transect across a sand dune.

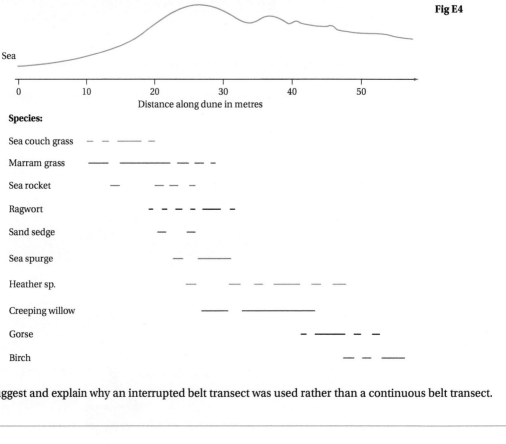

Fig E4

(a) Suggest and explain why an interrupted belt transect was used rather than a continuous belt transect.

_____

_____ 2 marks

(b) List three abiotic factors that make it difficult for plants to grow in sand dunes.

_____

_____

_____ 3 marks

(c) Name the pioneer species shown in the diagram.

_____ 1 mark

(d) Suggest two adaptations that allow these pioneer species to grow in sand.

_____

_____ 2 marks

Total marks: 8

6 Albinism in humans is caused by an autosomal recessive allele a. The dominant allele A gives normal colouration. A normal (non-albino) couple have three children: two are normal, one is albino.

    (a) What were the genotypes of the parents?

    _____ 1 mark

    (b) What is the probability that their next child will be an albino?

    _____ 1 mark

    (c) One of the normal children who carries the albino allele has a child with a partner who also has normal colouration. What predictions can be made about the colouration of their children? Explain your answer.

    _____

    _____

    _____ 2 marks

    (d) The albino child has a child with a partner of normal colouration. What predictions can be made about the colouration of their children? Explain your answer.

    _____

    _____

    _____ 2 marks

    Total marks: 6

7 In guinea pigs, black coat (B) is dominant to brown coat (b) and short hair (H) is dominant to long hair (h). These genes are autosomal and not linked.

    (a) Explain what is meant by:

        (i) Autosomal

    _____ 1 mark

        (ii) Not linked

    _____ 1 mark

  A guinea pig breeder has a pure-breeding, long-haired, brown male and a pure-breeding, short-haired, black female. A pet shop wants a supply of long-haired, black guinea pigs.

(b) Explain what the breeder will have to do to ensure a supply of pure-breeding, long-haired, black guinea pigs.

_____

_____

_____

_____

_____ 5 marks

Total marks: 7

8 Sickle-cell anaemia is a genetic disease in which abnormal haemoglobin causes sickle ('banana')-shaped red blood cells.

Normal homozygous individuals (SS) have normal blood cells that are easily infected with the malarial parasite. Thus, many of these individuals become very ill from the parasite and may die. Individuals who are homozygous for the sickle-cell trait (ss) have red blood cells that readily collapse into sickle shapes when deoxygenated. Although malaria cannot grow in these red blood cells, individuals often die because of the genetic defect. However, individuals with the heterozygous condition (Ss) have some sickling of red blood cells, but generally not enough to be fatal. In addition, malaria cannot reproduce within these 'partially defective' red blood cells. Thus, heterozygotes tend to survive better than either of the homozygous conditions.

If 9% of an African population is born with the severe form of sickle-cell anaemia (ss):

(a) Use the Hardy-Weinberg equation to calculate what percentage of the population is heterozygous (Ss) for the sickle-cell gene.

_____

_____

_____

_____ 4 marks

(b) Explain what would happen to the s allele in a population where malaria was eradicated.

_____

_____ 2 marks

Total marks: 6

9 **(a)** Explain the difference between a positive and a negative feedback.

_____

_____ 2 marks

Testosterone is the main sex steroid in males. The flow diagram (Fig E5) illustrates the control of testosterone.

(b) Explain how a negative feedback system operates to keep the levels of testosterone constant.

_____

_____

_____ 3 marks

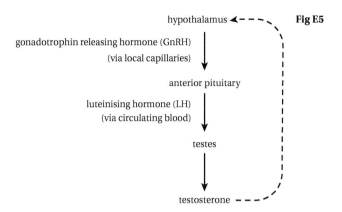

Fig E5

LH (Fig E5) is a hormone also found in females, where it has a different function.

(c) Explain how the same hormone can have different effects on males and females.

_____

_____ 2 marks

(d) Outline the mode of action of steroid hormones.

_____

_____

_____ 3 marks

Total marks: 10

10 One type of colour blindness is caused by a single gene located on the X chromosome. Normal individuals possess the allele C, which is dominant over the allele c which causes colour blindness. A female who carried the disease would have the genotype X^CX^c.

The pedigree diagram (Fig E6) shows three generations of the same family.

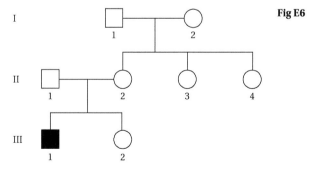

Fig E6

Key:
square = males
circle = females
black = colour blind
white = normal vision

(a) Give the genotype of individuals 1 and 2 in each of the three generations. Some individuals may have more than one possible genotype.

_____

_____

_____ 3 marks

(b) (i) The affected individual in generation 3 has a partner who has normal vision but is a carrier. What are the possible genotypes and phenotypes of their children? Show your working.

_____

_____

_____ 3 marks

(ii) What is the probability that the next child will be a male and colour blind?

_____ 1 mark

Total marks: 7

11 One line of stem cell research is focusing on creating new skin to help burn victims and people with diabetic skin ulcers. Stem cells are seeded onto a synthetic scaffold before being transferred to the wound. It is hoped that this new treatment will avoid the need for skin grafts.

(a) Explain what is meant by stem cell.

_____

_____ 2 marks

(b) Explain how a stem cell becomes specialised to form a skin cell.

_____

_____ 2 marks

(c) Explain why the new treatment is preferable to skin grafts.

_____

_____ 2 marks

(d) iPS cells are induced pluripotent cells.

(i) Explain the word induced.

_____

_____ 2 marks

(ii) Explain the word pluripotent.

_____

_____ 2 marks

(e) Give **two** possible advantages of iPS cells over embryonic stem cells in treating human disorders.

_____

_____ 2 marks

Total marks: 12

12 The disease GSD 1a (glycogen storage disease type 1a) results from a recessive allele that fails to make the enzyme glucose-6-phosphatase. As a result, individuals cannot break down glycogen when required. At present there is no cure, and sufferers have to eat large amounts of starch.

(a) Explain how a healthy individual responds to a fall in blood glucose levels (details of hormone action not required).

_____

_____

_____ 3 marks

(b) Explain why heterozygotes do not suffer from the disease GSD 1a.

_____

_____ 2 marks

(c) Explain why sufferers from GSD 1a have to eat large amounts of starch.

_____

_____ 2 marks

Some recent research has focused on the possibility of treating GSD 1a with gene therapy. Working copies of the glucose-6-oxidase gene are put into viruses that will transfer the gene into liver cells.

(d) Explain why viruses are suitable vectors in gene therapy.

_____

_____ 2 marks

(e) Suggest two possible problems with this type of gene therapy.

_____

_____ 2 marks

Total marks: 11

13 (a) Explain what is meant by the term *genome*.

_____

_____ 2 marks

(b) Give one advantage and one disadvantage of our knowledge of the human genome.

    (i) Advantage:

    _____

    _____ 1 mark

    (ii) Disadvantage:

    _____

    _____ 1 mark

Total marks: 4

14 Write an essay on **one** of the topics below.

EITHER

Biologically important compounds containing phosphorus. [25 marks]

OR

Negative feedback and its importance in biology. [25 marks]

Answers

Question	Answer	Marks
1 (a)	*Ethical – any one from*: Minimise damage to the organisms/ecosystem. Don't leave litter/rubbish. *Safety – any one from*: Wear suitable footwear. Avoid steep/slippery/sharp rocks. Avoid rough seas/deep water.	2, one for each
1 (b)	*Any three from*: Place line from low water line to splash zone (or words to that effect). Place transect on line. Record presence/absence or percentage cover of organisms. Move transect along/repeat at regular intervals.	3, one for each
1 (c)	A transect shows gradual change. Quadrats are better for comparing two different areas.	2
1 (d)	The transect gives qualitative data. Transect shows what species are present, *not* how many/relative abundance.	2
1 (e) (i)	*Any two from*: channelled, spiral, egg or bladder wrack; sugar kelp.	1, half for each
1 (e) (ii)	*Any two from*: wave action; time exposed to air/desiccation (drying out); temperature; salinity/water potential (refers to evaporation from rock pools).	2, one for each
		Total 12
2 (a)	Grazing will have no effect on species diversity.	1
2 (b)	*Any two from*: Map out the area/make grid/coordinates. Select coordinates at random. Use table of random values/computer program.	2, one for each
2 (c)	It is unlikely that they occurred by chance.	1
2 (d)	Chi squared.	1
2 (e)	Grazing lowers species diversity. Grazing has no effect on the grasses or buttercup. Grazing has a significant effect on dandelion/willow/bilberry/trefoil.	3, one for each
2 (f)	Increase in humus/nutrient content of soil; succession by more perennial/woody plants; greater diversity of plant and animal species; climax community of deciduous woodland; named dominant species, e.g. oak, beech, birch.	5
		Total 13
3 (a)	*Any two from*: Organisms take time to ingest food; they increase in volume/grow; they need to become mature enough to reproduce; time needed to activate genes; time taken to synthesize enzymes.	2, one for each
3 (b)	Growth was rapid in both species because there were *no limiting factors*. *Any two from*: plentiful food; no predation; no waste accumulation.	3

Question	Answer	Marks
3 (c)	*P. aurelia* has *out-competed P. caudatum*. *Possible explanations*: *P. aurelia* better at catching food –or– *P. aurelis* is producing a substance/toxin that inhibits *P. caudatum*.	2
		Total 7
4 (a)	The biggest fall in snowshoe hare numbers was around 1868/69.	1
4 (b)	*Any four from*: The higher the hare population, the greater the population of lynx that can be supported (*or vice-versa*). There is a *carrying capacity* – a certain population of lynxes can be supported by a certain population of hares. The more hares there are, the more food for the lynx and the more cubs they can rear. The lynx population is linked to the hare population. An increase in hare numbers is followed by an increase in lynx numbers ... ... after a lag/time delay.	4, one for each
4 (c)	Most predators eat more than one prey species. So interactions between populations are more complex.	2
		Total 7
5 (a)	Long distances are involved. A continuous transect would take too long.	2
5 (b)	*Any three from*: very little water; very few nutrients/minerals/humus; constantly shifting sand; plants can get buried; high salt content; high wind causes desiccation (drying out).	3
5 (c)	Marram grass and sea couch grass (*accept both or either one*).	1
5 (d)	*Any two from*: extensive roots consolidate sand; deep roots reach water; rolled leaves/sunken stomata/hairs reduce transpiration.	2, one for each
		Total 8
6 (a)	Both parents must be genotype Aa.	1
6 (b)	The probability is: 0.25, 25% or 1 in 4 (all mean the same).	1
6 (c)	If partner is AA, there is no chance that the children will be albino. If partner is also a carrier (i.e. Aa), there is a 1 in 4 chance.	2
6 (d)	If the new partner is AA, then all the children will have normal colouration. The children will be genotype Aa. If partner is Aa, there is a 1 in 2 (50%) chance that the children will be albino.	2
		Total 6
7 (a) (i)	Autosomal means: genes not located on sex chromosomes.	1
7 (a) (ii)	Not linked means: genes not located on same chromosomes.	1

Question	Answer	Marks
7 (b)	*Any five from*: First generation are all Bbh – Hh. Second generation – all possible genotypes. The breeder wants BBhh individuals. Long-haired, black guinea pigs might be Bbhh or BBhh. To find out which, do a test cross with a white-coated individual. If all offspring are all black, individual was BBhh – pure breeding. If half black, half white, individual was Bbhh – not pure breeding.	(1 mark given for correct Punnett square) 5, one for each
		Total 7
8 (a)	The frequency of ss (i.e. q^2) is 0.09. Therefore the frequency of the q allele is 0.3. $p + q = 1$; therefore the frequency of the p allele is 0.7. The frequency of the Ss genotype is $2pq$, i.e. $2 \times 0.7 \times 0.3 = 0.42$ or 42%.	4
8 (b)	*Any two from*: Ss individuals no longer have an advantage. SS individuals would have an advantage over Ss and ss. Frequency of the s allele would decrease.	2, one for each
		Total 6
9 (a)	Negative feedback is a mechanism that keeps conditions constant. Positive feedback is a mechanism that brings about further change.	2
9 (b)	Too much testosterone is detected by the hypothalamus. Therefore less GnRH is made. So less LH is made and less testosterone is made/secreted. (**NB:** *converse approach also acceptable – too little testosterone ...*)	3
9 (c)	*Any two from*: Males and females have different membrane receptor proteins. There is a different pathway of second messengers in the cytoplasm. Different enzymes are activated/deactivated.	2
9 (d)	*Any three from*: Hormone passes through cell membrane. It binds to nuclear/intracellular receptors. Transcription factors are activated. Specific genes are expressed/transcribed.	3
		Total 10
10 (a)	*1 mark for each generation all correct*: Generation 1 – Male $= X^CY$; Female $= X^CX^c$ Generation 2 – Male $= X^CY$; Female $= X^CX^c$ Generation 3 – Male $= X^cY$ (i.e. colour blind); Female $= X^CX^c$ or X^CX^C.	3
10 (b) (i)	*1 mark each (for parents, gametes and offspring)*: Parental genotypes - Male $= X^cY$; Female X^CX^c Gametes X^c, Y, X^C Offspring $= X^CX^c$ (female, normal vision) $\quad\quad\quad X^CY$ (male, normal vision) $\quad\quad\quad X^cY$ (male, colour blind)	3

Question	Answer	Marks
10 (b) (ii)	0.25 (0.25 is correct because probabilities are expressed in decimals. It means the same as 25% or 1 in 4 but the question doesn't ask for the answer in that form.)	1
		Total 7
11 (a)	A stem cell is an undifferentiated cell that has the potential to develop into other specialised cell types.	2
11 (b)	Specialisation is caused by the selective activation of genes. Particular genes are transcribed/expressed. This happens when all necessary transcription factors are in place.	2
11 (c)	Taking skin from another part of the body is painful. Reference to healing time for that part of body/infection risk.	2
11 (d) (i)	*Any two from:* iPS cells have been reprogrammed Using transcription factors So they are more potent The specialisation process has been reversed	2
11 (d) (ii)	The cells have the potential To differentiate into any cell type in the body	1 1
11 (e)	No embryos are destroyed. iPS cells are the patient's own cells, so less chance of rejection (because all cells have the same/similar combination of cell surface proteins).	1 1
		Total 12
12 (a)	It would be detected by α cells in islets of Langerhans which secrete glucagon. Glucagon activates enzymes/glucose-6-phosphatase. This stimulates breakdown of glycogen.	3
12 (b)	Heterozygotes possess one healthy allele. One healthy allele is all they need to make the working enzyme.	2
12 (c)	The digestion/hydrolysis of starch ... ... releases glucose slowly into the blood preventing glucose levels falling too low.	2
12 (d)	Viruses can be made to target specific cells. Thus deliver DNA/genes/genetic material directly into the cell.	2
12 (e)	*Any two from:* It is not possible to be absolutely certain what the virus will do in the body. It could enter cells other than the liver. It could cause disease. Immune reactions to the altered virus might cause problems.	2
		Total 11

Question	Answer	Marks
13 (a)	All of the DNA sequences in an organism.	1
	All the genes and all of the non-coding sequences in between.	1
13 (b) (i)	Advantage: (1 mark for a suitable example)	
	• Greater understanding of genes and how they interact.	
	• Potential to treat/cure genetic disease.	
	NB there will be a large range of acceptable answers for both parts of this question.	1
13 (b) (ii)	Disadvantage: (1 mark for a suitable example)	
	• Can lead to difficult early diagnosis of potentially fatal conditions.	
	• Can lead to prejudice against individuals with certain conditions.	
	• For example, refusal of job/life insurance/mortgage.	
	• Can lead to problems with companies trying to patent biological molecules/ gene products.	1
		Total 4
14	The marks are in five bands according to the level of response – see specimen mark schemes from the awarding organisation. In order to reach the highest mark bands students must also include at least **five topics** in their answer, to demonstrate a synoptic approach to the essay.	
	Essay 1	
	Biologically important compounds containing phosphorus.	
	Phospholipids	
	• Structure of the phospholipid molecule and its behaviour in water.	
	• Bilayers and micelles.	
	• Fluid mosaic theory.	
	Nucleic acids:	
	• DNA. Outline of structure.	
	• Role in replication.	
	• Role in protein synthesis.	
	• RNA – different types.	
	• The role of the different types in protein synthesis.	
	ATP	
	• Properties of the molecules: instant energy in one simple chemical step.	
	• Production of ATP in both respiration and photosynthesis.	
	Negative feedback and its importance in biology	
	Basic principles	
	• Features of a negative feedback (detection, correction, monitoring).	
	• Idea and importance of homeostasis.	
	Role of the nervous system	
	• The autonomic nervous system, e.g. in the control of heart rate and therefore the control of oxygen and carbon dioxide levels.	
	• Simple behaviour patterns. Taxes, kineses, tropisms to keep organisms in a favourable environment.	
	• Effects of exercise on circulation and breathing.	
	Role of the endocrine system/hormones	
	• Control of blood glucose (insulin and glucagon).	
	• Water balance (the role of ADH).	25

Glossary

Abiotic factor	Non-living environmental factor, such as temperature, pH, humidity, carbon dioxide level. Compare with **biotic factor**.
Acetylation	Addition of an acetyl group ($COCH_3$); one of several histone modifications that can control transcription.
Addition	Type of gene mutation in which a base is added, causing a frame shift in one direction, so that many codons are changed. See also **deletion**, **substitution**, **mutation**.
Adult stem cells	Multipotent cells found only in specific adult tissues, including bone marrow, that can only mature into a limited number of cell types.
Allele	An alternative form of a gene. For example, a flower colour gene could have an allele for red flowers and one for white flowers.
Allopatric speciation	Type of speciation in which a new species develops when physically separated from the original population. Compare with **sympatric speciation**.
Autosome	Any chromosome that is not a sex-determining chromosome.
Base	In nucleic acids (DNA and RNA), one of four nitrogen-containing compounds that fit together like jigsaw pieces. In DNA the bases are adenine, thymine, guanine and cytosine. RNA has uracil instead of thymine.
Benign	A non-cancerous tumour.
Biodiversity	Biological diversity; a measure of the number of species living in a certain area. Some habitats have a very high biodiversity (for example, rainforests, coral reef) and some a very low biodiversity (for example, polar regions and deserts). Reduction in biodiversity due to human activity is a major concern.
Biotic factor	Environmental factor caused by other organisms, such as food supply, predation or disease. Contrast with **abiotic factor**.
Blastocyst	Embryonic stage, before implantation, when the embryo consists of a hollow ball of cells.
Carcinogen	Cancer-causing agent.
Cardiomyocytes	Heart muscle cells, generated from unipotent cardiac progenitor cells.
Carrying capacity	The size of population that can be supported in any given situation. This could be the number of tigers in a certain area or forest, or the number of aphids on a plant. Two key factors that affect carrying capacity are usually food availability and predation.
Chi squared (χ^2)	Statistical test that compares observed values (i.e. those gathered in the investigation) with those you would expect if there were no correlation. Named after the Greek letter chi, χ.
Chiasmata	The points on a chromosome pair that remain in contact during the first stage of meiosis and at which crossing over and exchange of genetic material occur between the strands.
Chromosome	Highly condensed ('super-coiled') DNA molecule that appears in a cell just before cell division. The name means 'coloured body'.

Climax community	Situation where an ecosystem has matured and stabilised after a period of **succession**; for example, deciduous woodland, tropical rainforest. Characterised by a few dominant species, usually trees.
Cloning	Process of making a genetically identical copy. Can apply to a piece of DNA, a gene, cell, tissue or whole organism.
Codominance	A situation where neither allele is recessive and so if both are present in the genotype, both are expressed in the phenotype. Seen in type AB blood.
Codon	A group of three bases in DNA or RNA that codes for a particular amino acid. Also called a **triplet**.
Colonisation	In the development of ecosystems, colonisers are the first organisms to take hold in a non-living environment; for example, lichens on bare rock. Colonisers improve the abiotic environment so that a wider variety of organisms can live there, and succession begins.
Community	The living component of an ecosystem; i.e. the interacting populations of all the different species.
Competitive exclusion principle	Idea stating that no two species can occupy the same niche in an ecosystem.
Complementary DNA (cDNA)	Which has been made from mature RNA by reverse transcription. Unsurprisingly, catalysed by the enzyme reverse transcriptase.
Continuous belt transect	Sampling method used to show the change in species from one area to another. In a continuous belt transect, quadrats are placed alongside each other with no gaps.
Crossover	In meiosis, the process that swaps blocks of genes between homologous chromosomes. Creates new combinations of alleles, so increasing variation.
Deciduous forest	Forest in which most of the dominant tree species shed their leaves; for example, oak, ash, beech; the **climax community** in the UK.
Degrees of freedom	A measure of how many values can vary in a statistical calculation; equal to the number of values in a data set.
Deletion	Type of gene mutation in which a base is lost (deleted). This causes a **frame shift**.
Deme	An interbreeding population.
Density-dependent factor	Environment factor that depends on the size or density of a population. For example, the higher the population, the greater the competition for food.
Dideoxy sequencing	Technique used to find the base sequence of a piece of DNA.
Dihybrid cross	Cross involving two separate genes. Notation used is usually something like: AABB × AaBb.
Diploid	Cell or organism that possesses two sets of chromosomes. Often written as $2n$. In humans, $2n = 46$.
Directional selection	Type of natural selection that favours one extreme of phenotype, for example tallest, quickest, heaviest.
Disruptive selection	Type of natural selection that favours both extremes of phenotype over the mid range.
Diversity	Measurement of the number of different species present in an ecosystem. The diversity of different ecosystems, or change over time, can be measured with a diversity index.

DNA polymerase	Enzyme that catalyses the addition of complementary nucleotides during DNA replication.
DNA probe	Marker that is attached to a fluorescent or radioactive label, used to find a specific sequence of nucleotides in a DNA molecule. The probe contains pieces of single-stranded DNA complementary to a certain base sequence so that the probe attaches to this specific sequence when the probe is added to a DNA sample.
Dominant	An allele which, if present, is shown in the phenotype.
Duplication	Type of mutation in which one or more copies of any piece of DNA including a gene or even an entire chromosome is produced.
Ecosystem	Natural unit such as a lake, woodland, coral reef, etc. that contains many different species together with the non-living components.
Electrophoresis	A technique which separates molecules in a mixture according to their size or charge. The most well-known use is the separation of differently sized DNA fragments – a vital part of DNA profiling.
Environmental resistance	The inevitable restriction of population growth, which always slows down sooner or later due to limiting factors.
Epigenetics	The effect of environmental factors on gene expression, resulting in heritable changes in gene function.
Epistasis	Occurs when one gene's allele masks the phenotypic effects of another gene's allele.
Exponential phase	In population growth, stage of rapid growth due to a lack of limiting factors, i.e. when conditions are favourable. See also **logarithmic phase.**
Expressed	A gene is expressed when it is active and being used to make a particular protein/polypeptide.
Factor VIII	Protein involved in blood clotting. Inability to make factor VIII is the cause of one type of **haemophilia**.
Frame shift	In mutation, a situation where a base is added or lost, causing all other bases to move along one place in a particular direction.
Gamete	A sex cell: sperm in males, eggs in females.
Gene	A length of DNA that codes for the production of a particular polypeptide or protein.
Gene pool	The sum total of the **alleles** circulating in an interbreeding population, or **deme**.
Gene therapy	The treatment of genetic disease by replacing defective alleles with functional alleles.
Genetic fingerprinting	The analysis of DNA fragments that have been cloned by PCR to identify an individual or to determine genetic relationships between organisms.
Genome	The entirety of the DNA base sequences in an organism. Includes all of the genes on all of the chromosomes, and the entire non-coding DNA between the genes. The human genome consists of 23 chromosomes, about 21 000 genes and just over 3 000 000 000 (3 billion) bases.
Genotype	The alleles that an organism has. Aa or AA, AaBb or Aabb and so on. Compare with **phenotype**.

Habitat	The external environment in which an organism lives.
Haemophilia	Sex-linked genetic disease in which blood fails to clot properly due to an inability to make **factor VIII**. Caused by a recessive, sex-linked allele. Males can't be carriers, so if they inherit the allele, they have the disease.
Hardy-Weinberg	Law which states that in a large, randomly mating population, allele frequencies will remain constant from one generation to the next, unless there is mutation, immigration, emigration or selection.
Heterozygous	Possessing two different alleles. Written as Aa or Bb.
Histone	Class of protein that organises the DNA in the nucleus. DNA winds round a histone, like cotton round a bobbin.
Homozygous	Possessing two alleles the same, for example, AA or aa. Said to be 'true breeding'.
In vitro	'In glass' – processes carried out in test tubes, etc.
In vivo	'In life' – processes happening in living cells/tissues or whole organisms.
Induced pluripotent stem (iPS) cells	Stem cells produced from adult specialised (unipotent) cells, reprogrammed to behave like embryonic stem cells.
Inner cell mass	Cluster of pluripotent cells inside the hollow sphere of a blastocyst.
Interrupted belt transect	Sampling method used to show the change in species from one area to another. In an interrupted belt, quadrats are placed at intervals along the transect. Suitable for a longer transect where the change is gradual. Compare with **continuous belt transect**.
Interspecific competition	Competition between individuals of *different* species.
Intraspecific competition	Competition between individuals of the *same* species.
Introns	Non-coding DNA within a gene. Introns must be removed before translation.
Inversion	Type of gene mutation in which the sequence of nucleotide bases on part of a chromosome is reversed.
Isolation	Vital step in speciation. Two populations become isolated when they cannot interbreed. Natural selection will then act in different ways on the isolated populations.
Lag phase	In population growth, an initial period of slow growth.
Ligase	Enzyme that joins together complementary sticky ends of DNA. Used in conjunction with restriction enzymes.
Limiting factor	The factor that is in short supply. If supply is increased, the rate of the process increases too. Light is a common limiting factor for photosynthesis.
Lincoln index	Formula for estimating population size using the **mark-release-recapture** method.
Linkage	The property of genes of being inherited together with other genes located on the same chromosome.
Locus	Position of a gene/allele on a chromosome.
Logarithmic phase	In population growth, period of rapid growth because there are no limiting factors. Also known as **exponential phase.**
Malignant	A cancerous tumour. Tends to grow at the edges and invade surrounding tissue.

Marker gene	Type of gene, for example an antibiotic resistance gene, used to determine which cells have been transformed by taking in the vector (genetically modified plasmids).
Mark-release-recapture	Method for estimating the population of a particular animal species. The animal is trapped, harmlessly marked and then released. On a second occasion, the population size can be estimated from the number in the second sample that are already marked. See **Lincoln index.**
Meiosis	Cell division that shuffles the genes on the chromosomes so that no two gametes are the same. One diploid cell gives rise to four haploid cells.
Mendelian population	An interbreeding population. Also called a **deme.**
Messenger RNA (mRNA)	A single strand of nucleotides that is made on a gene during **transcription.** Basically, it is a mobile copy of a gene.
Metastasis	Process in which cancerous cells break off from a malignant tumour and set up secondary tumours elsewhere in the body.
Methylation	Addition of a methyl group (CH_3) to DNA; an important method of epigenetic regulation.
Microhabitat	The environmental conditions in an organism's immediate surroundings; for example, the conditions under a stone on a river bed will be completely different from those above it – light, temperature, oxygen levels, flow rate and risk of predation will all be different.
Monohybrid cross	A cross involving a single gene.
Multiple alleles	Refers to a gene that has more than two alleles. Many genes have one allele, because they code for a vital product, such as an enzyme, and any mutations are lethal. However, some genes have two alleles and some have more than two. The gene for the human blood group is a good example of multiple alleles: there are three alleles: I^A, I^B and I^O that combine to produce the blood groups A, B, AB and O.
Multipotent	Multipotent stem cells have limited potential, and can only differentiate into a few closely related cell types.
Mutation	A change in an organism's DNA. Gene mutations are changes in the base sequence of a particular gene, while chromosome mutations involve changes in whole blocks of genes.
Natural selection	Differential survival of organisms; having features that best adapt organisms to their environment, making them more likely to survive and reproduce.
Niche	Concept that describes an organism's position in an ecosystem.
Nucleotide	Basic sub-unit of a nucleic acid, consisting of a sugar (deoxyribose or ribose), a phosphate and one of four bases.
Null hypothesis	A hypothesis that takes a negative position on an investigation, such as 'there is no link between cancer and smoking' so that it can be disproved/refuted by statistics. It is the nature of scientific investigation that you can never prove anything, but you can disprove things. By testing the null hypothesis with statistics, you can reject it and so gather support for your hypothesis. See **chi-squared.**
Oestrogen	A steroid hormone, made in the ovaries, which has a vital role in the menstrual cycle. Being lipid-soluble, the oestrogen molecule can enter target cells. It combines with oestrogen receptors in the cytoplasm and then passes into the nucleus, where it acts as a transcription factor for many different genes.

Oncogene	A gene that causes cancer. Normal, **proto-oncogenes** control cell division. If they mutate, they become oncogenes. See also **tumour suppressor genes**.
Palindromic sequence	A section of DNA where the sequence of bases on one strand reads the same as the sequence on the complementary strand when read from the other direction. For example, CGTACG will read GCATGC on the complementary strand, which is CGTACG when read from the other direction.
Phenotype	The observable features of an organism (genotype + environment = phenotype). Compare with **genotype**.
Phenotypic ratio	The proportion of one phenotype to another, expected from a genetic cross.
Pioneer species	In the development of ecosystems, pioneer species are the first to establish themselves. They can usually tolerate harsh conditions and low nutrient levels. Examples include lichen on bare rock, or marram grass on sand dunes. Also called **colonisers**.
Plasmid	Tiny circles of DNA found in the cytoplasm of bacteria. Contain useful rather than essential genes.
Pluripotent	Type of stem cell that can differentiate into a wide variety of the 216 different cell types. Has some potential in gene therapy.
Point transect	**Transect** in which the individual species touching a particular point on the line (for example, every 10 cm) are recorded.
Polymerase chain reaction (PCR)	A method of cloning DNA in a test tube. This process requires an original sample of DNA, nucleotides, primers and the enzyme DNA polymerase.
Population	Group of individuals of the same species that can interbreed.
Predation	When one organism (the predator) eats another (the prey). An important **biotic factor**. The numbers of predator and prey are often inter-dependent.
Promoter region	Part of a gene that initiates transcription. The promotor is usually placed 'before' or 'upstream' of the actual coding part of the gene. The RNA polymerase enzyme attaches to the promotor region. Transcription will not begin until all the necessary transcription factors are in place. Once a complete Transcription Initiation Complex (TIC) has formed, the RNA polymerase enzyme will transcribe the gene, making a molecule of mRNA.
Proteome	The full range of proteins that an organism can make, as encoded by the genome.
Proto-oncogenes	Proto-oncogenes code for proteins that aid the regulation of differentiation and cell growth. (A normal gene that can become an oncogene from mutations or increased gene expression.)
Punnett square	In genetics, a grid for organising all the possible outcomes of a cross.
Quadrat	In ecology field work, a sampling method used to compare two different areas. Often a small, 0.5 m square frame.
Qualitative data	Data that show 'what' rather than 'how much'.
Quantitative data	Data that show 'how much' as well as 'what'.
Recessive	An allele that only appears in the phenotype if the dominant allele is absent. For example, *aa*.

Recognition site	Sequence of bases on a DNA molecule that is cut by a restriction enzyme, producing sticky ends. Recognition sites are usually palindromic, i.e. the sequence on opposing strands is the same, such as CTATAG, whose **palindromic sequence** is GATATC.
Recombinant DNA	DNA molecules formed by joining together (recombining) DNA from two different sources (usually different organisms). See also transgenic organisms.
Reproductive success	The ability of an organism to pass its alleles/allele combinations on to the next generation. Sometimes referred to as 'fitness' and often simply measured by the number of offspring.
Restriction endonuclease	Enzyme that cuts DNA at specific recognition sites, to produce **sticky ends**.
Restriction site	Location on a DNA molecule containing specific sequences of nucleotides, which are recognised by restriction enzymes.
Reverse transcriptase	Enzyme that makes DNA from RNA: **transcription** in reverse, hence the name.
RNA interference (RNAi)	A process that inhibits translation of the mRNA produced from target genes in eukaryotes and some prokaryotes.
RNA polymerase	Enzyme that catalyses the addition of complementary nucleotides during transcription. The first stage of protein synthesis where the messenger RNA (mRNA) is assembled on a gene.
Sampling	When gathering data, it is not necessary (and usually not possible) to take every measurement. The idea of sampling is to get accurate and reliable data without taking up too much time. For example, if you want to get an accurate idea of the weight of all the voles in a woodland, you would catch and weigh individual voles, working out a running mean as you go. When the mean stops changing significantly, you've taken enough measurements and taking any more won't alter the mean.
Screening	Examination or testing of a group of individuals to separate those who are well from those who have an undiagnosed disease or defect or who are at high risk.
Short interfering RNA (siRNA)	Small RNA molecules that can block transcription. Have potential in medicine to block harmful genes.
Speciation	The process of forming a new species from pre-existing species.
Stabilising selection	Type of natural selection in which middle values have a selective advantage.
Stem cell	Cell with the ability to differentiate into various specialised cells.
Sticky ends	Staggered cuts in DNA, so that one strand is several bases longer than the other. Made by restriction enzymes.
Substitution	Type of mutation in which a particular base is replaced by a different base. Also known as a point mutation. Changes one codon, and one amino acid, but still has the potential to change the whole protein. Contrast with **addition** and **deletion**.
Succession	The development sequence of an ecosystem, in which one set of plant species changes the abiotic environment so that a different set of species takes over. Sequence continues until a **climax community** is established.
Sustainable	In agriculture, food production that can be maintained year on year without a depletion of natural resources or permanent damage to the ecosystem.

Sympatric speciation	One type of speciation, where isolation takes place even though the two populations live together. Clear examples are rare. Compare with **allopatric speciation**.
Taq polymerase	Thermostable polymerase enzyme from the bacterium *Thermus aquatalis*. Used extensively in PCR because it is not denatured at the high temperatures needed.
Terminator region	Part of a gene that terminates transcription. The terminator is found 'downstream' of the coding part of the gene, and signals the RNA polymerase enzyme to stop transcribing. As a consequence, the mRNA molecule will detach from the DNA strand.
Totipotent	Type of stem cell, found in the embryo, that is totally potent; i.e. has the ability to develop into any cell type.
Transcription Initiation Complex (TIC)	Formed when transcription factors bind with RNA polymerase, allowing RNA polymerase to bind to the site where transcription should start, and begin RNA synthesis.
Transcription	First stage (of two) in protein synthesis. The base sequence on a particular gene is copied onto a molecule of **messenger RNA (mRNA)**. Takes place in the nucleus.
Transect	In ecology field work, a sampling method in which the organisms present along a line are recorded to show change from one area to another.
Transformation	Inserting DNA into a cell or plasmid.
Transgenic organisms	Organisms that have had DNA from another individual – often from another species – inserted into their genome. Transgenic organisms therefore contain recombinant DNA.
Translation	Second stage (of two) in protein synthesis. The base sequence on the messenger RNA (mRNA) molecule is used to assemble a protein. Takes place on *ribosomes*.
Translocation	Type of gene mutation in which the sequence of nucleotide bases on part of a gene is removed and inserted elsewhere. Usually both the original gene with a missing section and the new gene with an attached section make non-functional proteins.
Triplet	Group of three bases. Used interchangeably with the word **codon**. A sequence of three bases (for example, AAT, CGC) that codes for a particular amino acid.
Tumour	A swelling caused when cells divide out of control. See also **benign** and **malignant**.
Tumour suppressor gene	Gene coding for a protein that prevents tumour formation. Acts as backup if the proto-oncogenes mutate. If the tumour suppressor genes also mutate, a tumour is much more likely to develop.
Unipotent cells	Cells that can differentiate into just one cell type.
Variable number tandem repeats (VNTRs)	A location in a genome where a short nucleotide sequence is organised as a tandem repeat. These can be found on many chromosomes and often show variation in length between individuals.
Vectors	Carriers. In genetic engineering, vectors carry genes or pieces of DNA from one cell to another. Liposomes and viruses are commonly used as vectors.
Zygote	A diploid cell resulting from the fusion of two haploid gametes; a fertilised ovum.

Index